GROUND VIBRATION ENGINEERING

GEOTECHNICAL, GEOLOGICAL AND EARTHQUAKE ENGINEERING

Volume 12

Series Editor

Atilla Ansal, *Kandilli Observatory and Earthquake Research Institute, Boğaziçi University, Istanbul, Turkey*

For further volumes:
http://www.springer.com/series/6011

Ground Vibration Engineering

Simplified Analyses with Case Studies and Examples

by

MILUTIN SRBULOV

United Kingdom

with Foreword of

E.T.R. Dean

 Springer

Dr. Milutin Srbulov
United Kingdom
srbulov@aol.com

Additional material to this book can be downloaded from http://extras.springer.com

ISSN 1573-6059
ISBN 978-90-481-9081-2   e-ISBN 978-90-481-9082-9
DOI 10.1007/978-90-481-9082-9
Springer Dordrecht Heidelberg London New York
100625608 X
Library of Congress Control Number: 2010929681

Printed on acid-free paper

Springer is part of Springer Science+Business Media (www.springer.com)

Foreword

Ground vibration is a significant geohazard and has always been an important consideration in civil engineering design and construction. Vibrations can be caused naturally, by earthquakes, landslides, wave action, volcanoes, river waterfalls, and other actions (Huang et al., 2007). Vibrations can also be caused by people and technology, such as by road vehicles, trains, shipping, cargo handling, industrial machinery, and war (Sheng et al., 2006). Ground vibration can be caused by very loud noises, including by musical events, and by people marching or dancing. It can be caused by construction plant and operations.

Levels of ground vibration that are strong enough to be felt by people are also strong enough to potentially damage structures and foundations. Damages can include excessive building settlement, liquefaction of sandy soils, slope instability, collapse of trenches, excavations, and tunnels, exposure of buried pipelines and other services, cracking of pipes, and serious discomfort to persons in workplaces, at home, or en route. Conversely, some construction operations deliberately vibrate the soil, such as vibro-coring methods. Other operations have ground vibration as an inevitable consequence, such as pile driving (White et al., 2002).

Ground vibration is also a useful tool in exploration, and can be used to find buried minerals and objects of many kinds. Sometimes things go wrong, however. Jefferies and Been (2006) tell the story of a number of exploration trucks which purposely sent sound waves into the ground, as part of a survey for hydrocarbons. The sound waves liquefied the sandy ground on which the trucks were parked, leading to a slope failure and to the trucks sinking into the quicksand that they had created.

This book brings together the main aspects of the practical knowledge of the art and science of ground vibration engineering. It describes techniques of measuring ground vibration, how to predict the likely maximum ground vibration levels during a structure's design life, how to predict its effects, how to design robust structures and foundations that will withstand design vibration levels, how to design construction operations that use ground vibration effectively when needed, how to minimize it when not, and how to conduct a forensic investigation into ground vibration damage.

Ground vibration is one of the expanding fields within civil engineering, and the practical knowledge and experiences recounted herein will likely help to encourage much needed further research. Some of the topics of current and future work are likely to include development of understanding of cyclic loading effects on soils, development of methods to accurately predict changes of soil stiffness and strength due to cyclic and dynamic loading, further development of safe geophysical methods of ground investigation, increased use of methods of controlled experimentation such as centrifuge modelling (Itoh et al., 2005), and further development of standards, codes of practice, and legislation.

As well as distilling the Author's extensive experience, this book provides extensive reference to and commentaries on technical standards and codes of practice, and references to the key practical and academic references in the technical literature.

Soil Models Limited, Aberdeen, UK   E.T.R. Dean
Caribbean Geotechnical Design Limited, Curepe, Trinidad

Preface

Ground vibration consideration is gaining significance with decreasing people's tolerance of vibration, introduction of new environmental legislations, increasing use of equipment sensitive to vibration, ageing of existing buildings and expanding construction sites to/near collapsible/liquefiable/thixotropic soil.

This volume bridges the gap that exists between rather limited provisions of engineering codes/standards and complex numerical analyses/small-scale tests. While a number of the codes/standards and text books are mainly concerned with the effects of vibration on humans and structures very few of them deal with vibration induced ground failures. Numerical analyses, even in elastic domain, require expert knowledge, which is available only within large/specialised companies and at universities.

This volume contains descriptions of ground vibration measurements, predictions and control for engineers. Effects of most frequent sources of ground vibration arising from construction/demolition, traffic and machinery have been considered by simplified analyses aimed at ease and speed of use for major problems in ground vibration engineering. Comments on assumptions, limitations, and factors affecting the results are given. Case studies and examples worldwide are included to illustrate the accuracy and usefulness of simplified methods. A list of references is provided for further considerations, if desired. Microsoft Excel spreadsheets referred to in Appendices and provided on http://extras.springer.com are for the case studies and examples considered in this volume.

Specialists in non-linear dynamics analyses recognize that the motion of a non-linear system can be chaotic and the outcomes can be unrepeatable and unpredictable. The non-linearity arises when stress-strain relationship is non-linear even in elastic strain range and when cracking and plastification occur on yielding of materials at large strain. Baker and Gollub (1992), for example, show that two conditions are sufficient to give rise to the possibility of chaotic motion: the system has at least three independent variables, and the variables are coupled by non-linear relations. Equivalent linear and simplified non-linear dynamic analysis described in this volume can be used to avoid possible chaotic outcomes of complex non-linear dynamic analyses/small-scale tests.

United Kingdom   Milutin Srbulov

Acknowledgments

I was honoured and privileged to work with Professor Nicholas Ambraseys as his assistant (i.e. associate after the completion of my PhD in 1994) on a number of research projects supported by the Engineering and Physical Science Research Council of the United Kingdom and by the EPOCH program of the Community of European Countries at Imperial College in London during the period 1992–1997. The simplified approach used in our research is directly applicable to routine engineering practice. My initial involvement with earthquake engineering was a starting point for further work concerning soil dynamics and ground vibration related issues while working on a number of projects worldwide.

Dr E.T.R. Dean reviewed several of my papers and was of great help with his detailed and precise comments for the improvement of initial versions of the papers. He kindly reviewed the initial version of this volume and made a significant contribution towards the improvement of the clarity and readability of the text.

The editor of European Earthquake Engineering journal Professor Duilio Benedetti accepted kindly for publication eighteen my papers related to geotechnical earthquake engineering in the period 1995 to end of 2007. Numerous publications of my papers encourage me to work more in the field of geotechnical earthquake engineering and related topics such as ground vibration.

The idea for writing books came from my wife Radmila, who had no time to do it herself but, nevertheless, provided encouragement and stimulation.

Contents

List of Symbols

Symbol	Description
$\bar{v}$	an average velocity through an equivalent homogeneous isotropic ground
ϕ_m	phase angle of a Fourier series
∞	symbol for infinity
$u_i, \dot{u}_i, \ddot{u}_i$	horizontal displacement, velocity and acceleration of a SDOFO at time instant i
$(N_1)_{60}$	normalized blow count N_{SPT}
$a(t), a[n]$	continuous in time and discrete at time intervals vibration amplitude
A_f	measuring instrument amplitude range, foundation area
A_{loop}	the area of the hysteretic loop
a_m, b_m, c_m	coefficients of mth member of a Fourier series
A_r	nominal amplitude of the vibrating roller
$A_{R,T,I}$	amplitudes of reflected (R), transmitted (T) and incident (I) waves
b	width of a slice of a cylindrical trial slip surface
b_c, l_c	breadth and length of a rectangular foundation or a pile cap
B_f	diameter of an equivalent circular foundation
$c(a_o)$	damping coefficient
c, c_o	viscous damping coefficient
c'	effective soil cohesion
$c_{t,p}$	ground wave velocity (t) transversal (p) longitudinal
$c_{u(1)}$	ground cohesion, (1) in one cycle i.e. static condition
$c_{u,avr}$	an average ground cohesion in undrained condition along pile shaft length
D_b	blast distance
d_h	the horizontal distance between the location where the load V is acting
d_i	thickness of a ground layer
D_p	external pile diameter
d_p	thickness of wall of a hollow pile
D_r	relative density of soil
dS	surface area
D_s	vibration source depth
$d\tau$	shear stress increment
E	Young modulus, energy
e	void ratio of soil
E_D	dissipated energy
E_{ff}	theoretical free-fall energy
E_{flux}	total energy-flux density per unit time
E_m	actual energy delivered by hammer

Symbol	Description
E_{max}	the maximum strain energy
$E_{o(r)}$	energy at vibration source (o) or at distance (r)
f	frequency of vibration
F	amplitude of a horizontal force acting on a foundation block
$f(t)$	Fourier series
$F_{A,V,D}$	Fourier amplitudes of acceleration (A), velocity (V), displacement (D)
f_c	corner frequency
f_d (f_o)	frequency of an input (output) motion
$F_d(\omega)$, $F[k]$	discrete Fourier transform
f_{max}	cut-off frequency
$F_{R,I}(\omega)$	real and imaginary part of a Fourier transform
F_s	factor of safety of slope stability
$F_{v,b}$	vertical foundation capacity
G	ground shear modulus
G_{max}	the maximum shear modulus
G_s	specific gravity of soil solids
G_{secant}	the secant shear modulus
H	thickness of a homogeneous ground layer
H_b	height of a building
H_d	the drop height
H_f	foundation (structure) height
H_l	distance from foundation level to the level of liquefied layer below
i	imaginary number
I, $I_{m,i}$	mass moment of inertia at a place i
I_x	second moment of cross section area at place x
$k(a_o)$	spring coefficient
K_e	static stiffness coefficient
$K_{h,v}$	stiffness of rubber bearings in the horizontal (h)/vertical (v) direction
K_o	the coefficient of soil lateral effective stress at rest
$k_{o,s,x,j}$	stiffness (force per displacement) (o) of an elastic damper, (s) structure, (x) soil, (j) at a place j
K_s	coefficient of lateral effective stress acting on pile shaft, coefficient of punching shear
K_θ	rotational stiffness of a foundation block
L_b	length of a beam
L_f	foundation length
L_p	length of pile in ground
L_s	vibration source length
L_w	wave length
m	mass of a foundation block
m, n, k	counters of a member of Fourier series
M_d	tamper weight
$m_{o,s,x,i}$	mass (o) of a SDOFO, (s) structure, (x) beam (i) at a location
N	even number of member in a fast Fourier transform
n	soil porosity
n_b	analogue to digital converter bit range
N_c	bearing capacity factor
n_d	number of vibration drums
N_i	average blow count of standard penetration test
n_m	vibration mode number

Symbol	Description
N_q	ground bearing capacity factor
N_{SPT}	blow count of standard penetration test (SPT)
N_x	axial force in a beam at place x
OCR	over consolidation ratio
p'_o	effective overburden stress at the foundation depth
P_f	maximum amplitude of vibration force at the ground surface
PI	soil plasticity index
P_Δ	reaction forces at the fixed ends of a beam displaced Δ_P at the end
r	resolution of instrument for ground vibration measurement
$r_{e,v,h,r}$	radius of equivalent disk (v) vertical motion (h) horizontal motion (r) rocking
r_f	radius of near field on the surface
r_o	radius of an equivalent disk
$r_{s,i}$	slant distance
S_e	dynamic stiffness coefficient
$S(\omega)$	power spectral density function
t	time
T_f	time interval of a harmonic function $f(t)$ defined in time t
T_n	period of the nth mode of free vibration of an infinite layer
T_p	period of small amplitude pendulum swaying
T_r	vibration record duration
T_w	wave period
u	horizontal displacement of a foundation block
$u^f(\omega)$	surface amplitude of the free field ground motion
u_w	pore water pressure at the base of a slice
V	vertical load acting at the foundation underside
$v_{(l,t,r)}$	ground wave velocity; l – longitudinal wave, t – transversal wave, r - Reyleigh wave
W	nominal energy of impact hammer
W_c	nominal energy of vibratory drivers per cycle
w_d	width of the vibrating drum
W_e	explosive mass
W_s	weight of a slice, vibration source width
$X[k], Y[k], F[k]$	coefficients of a fast Fourier transform
$x[n], y[n]$	odd and even members of $a[n]$
x_i	distance from impact
x_r	distance along the ground surface from the roller
z	Depth
Z_t	beam end displacement in time
ρ	unit density
$B, \Gamma, \beta, \beta_t$	tuning ratio
Δ_P	end displacement of a beam with fixed ends
Δt	time interval of a time series
$\partial \Delta_{w,l(t)} (\partial t)^{-1}$	ground particle velocity in longitudinal (l) and transversal (t) direction
$\Delta_{wr(o)}$	amplitude of displacement at distance (r) or at vibration source (o)
$\Delta \sigma_{v,z}$	vertical axial stress in soil at depth z
Ω_o	Fourier maximum displacement amplitude
$\Psi_{x,i,j}$	shape function of beam deflection (i,j) at a place i,j
α	angle of wave inclination to layer boundary or normal to the boundary
α_p	ground cohesion mobilization factor along pile shaft
α_s	inclination to the horizontal of the base of a slice

Symbol	Description
δ	small value of displacement, velocity, thickness etc.
δ_ϕ	friction angle between ground and pile shaft
ε_l	axial strain in longitudinal direction
ϕ	soil friction angle
γ	soil unit weight, shear strain
γ_d	dry unit weight of soil
γ_w	unit weight of water
κ	material attenuation coefficient
λ	wave length
ν	Poisson's ratio
θ	phase angle, angle of rotation of a foundation block
$\sigma'_{v(avr)}$	effective overburden pressure (avr, averaged over a depth)
$\sigma_{I,R,T}$	axial incident (I), reflected (R), transmitted (T) stress
σ_l	axial stress in longitudinal direction
σ_t	axial (tensile) stress
$\omega_{d,e,s,h,r}$	circular (angular) frequency (d) dumped, (e) equivalent, (s) structure, (h) in horizontal direction, (r) in rocking
$\xi_{e,g,s,h,r}$	damping ratio (e) equivalent (g) ground (s) structure (h) horizontal (r) rocking

Chapter 1
Problem Description

1.1 Introduction

Ground vibration consideration is gaining significance with decreasing level of people tolerance of vibration, introduction of new environmental legislations, increasing use of equipment sensitive to vibration, ageing of existing buildings and expanding construction sites to/near collapsible/liquefiable/thixotropic soil.

Vibration consideration involves its source, propagation path and its recipient. The objective of this chapter is to describe frequent sources of ground vibration, main vibration propagation media effects, sensitivities of recipients and legislation requirements. There are many sources of ground vibrations that can cause problems including vibrations from wave impact on shorelines, marching humans, movements of animal herds, landslides, waterfalls, thunder, loud sound waves, crowd cheering/stamping/Mexican waving, ship impacts on quay walls, swing bridge opening/closing, earthquakes, which effects are not topic of this volume. Ground vibration can be beneficial for compaction of sandy soil, for example.

Staring point of vibration consideration should be definition of acceptable vibration effect on a recipient.

1.2 Sensitivities of Recipients and Legislation Requirements

1.2.1 Humans

The effects of vibration on humans vary from annoyance to serious illness. Annoyance is subjective to different individuals and can be caused even by noise of vibrating structural elements instead directly by vibration. Many people are familiar with the noise of vibrating walls and floors caused by hand drills in their homes. Some environmental regulations consider effects from noise and vibration together. Noise and vibration can influence working efficiency by inducing stress, by disturbing concentration/rest and conversation at home and by increasing accident risk. The use of vibrating equipment such as pneumatic hammer when removing pavements

M. Srbulov, *Ground Vibration Engineering*, Geotechnical, Geological, and Earthquake Engineering 12, DOI 10.1007/978-90-481-9082-9_1,
© Springer Science+Business Media B.V. 2010

on streets, for example, can cause serious illness named 'vibration white fingers' to the operative. Health and safety at work regulations deal with such type of vibration problem, its avoidance and protection from it.

A number of subjective and objective factors influence human perception and reaction to ground borne vibrations, such as:

- Available previous information on type and duration of vibration i.e. its expectation. For example, some people will complain that noise and vibration in their homes causes disturbance of sleep while be prepared to sleep in airplanes, busses and cars, which cause equal levels of noise and vibration. Good public relations can help in decreasing the number of complains caused by ground vibrations.
- People age (youngsters are more sensitive), sex (pregnant woman are more sensitive), health state (people with neck or back problems, recently undergone any form of surgery and with prosthetic devices) (e.g. HSE, 2005).
- Worries about: property damage and reduction of its value/costs for its repair, damage to vibration sensitive equipment, performing sensitive tasks in operating theatre, precision laboratories, etc.
- Location of place of vibration such as residential, office, workshop, hospital, laboratory.
- Time of day i.e. day or night.
- Duration of vibration such as temporal due to construction or permanent due to a new traffic route, etc.

A number of publications provide information on the effects of different vibration intensities on humans (e.g. BS 5228-2, 2009). Also, standards/codes provide recommendations on maximum peak particle velocities/peak accelerations for perception/tolerance of humans (e.g. ANSI S 3.18, 1979; ANSI S3.29, 1983; BS 6472, 1992). Similarly, health and safety regulations provide requirements on maximum exposure to vibration of workers and the ways of their protection (e.g. HSE, 2005). A rather comprehensive overview of codes and standards related to ground vibrations is provided by Skipp (1998), New (1986), Hiller and Crabb (2000), etc.

1.2.1.1 Example of Guidelines in Codes for Vibration Limits Acceptable to Humans in Buildings

Both ANSI S3.29 (1983) and BS 6472 (1992) recommend the same basic root mean square (r.m.s.) accelerations in the vertical direction for critical working areas such as hospital operating theatres and precision laboratories shown in Fig. 1.1. The r.m.s. acceleration is the square root of the average of sum of squares of componential accelerations. Both codes recommend the multiplication factor of 4 of the basic r.m.s. acceleration for offices, and 8 for workshops for continuous (and intermittent vibrations and repeated impulsive shock according to ANSI S3.29, 1983) and 128 for both offices and workshops for impulsive vibration excitation (with duration less than 2 s) with up to 3 occurrences a day. These two codes differ only concerning

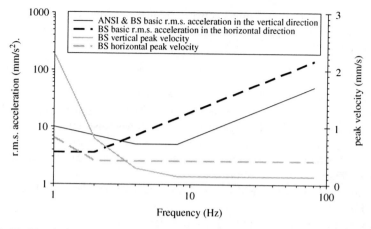

Fig. 1.1 Limiting for humans basic root mean square accelerations and compontential peak velocities for the vertical and horizontal directions versus frequency of vibration in buildings according to ANSI S3.29 (1983) and BS 6472 (1992)

Table 1.1 Multiplication factors of the basic r.m.s. acceleration in residential buildings

| Time | Continuous vibration | | Impulsive vibration (duration < 2 s) with up to 3 occurrences | |
	ANSI S3.29 (1983)	BS 6472 (1992)	ANSI S3.29 (1983)	BS 6472 (1992)
Day (7–22 h)	1.4–4	2–4	90	60–90
Night (22–7 h)	1–1.4	1.4	1.4	20

the multiplication factors of the basic r.m.s. accelerations for residential buildings as shown in Table 1.1. In addition, BS 6472 (1992) recommends the use of the same multiplication factors for the peak velocity.

1.2.2 Equipment

Excessive vibration can cause malfunction and damage of sensitive equipment. Manufacturers of equipments specify tolerable levels of vibrations for their equipment. In order to compare some of these levels with the acceptable levels of vibration for humans and structures, the following list is provided for example from Dowding (2000).

- IBM 3380 hard disk drive: 18 mm/s between frequencies from 1 to 200 Hz (0.3 g in the vertical direction, 0.1 g for 5 Hz, 0.3 g for 16 Hz, and 0.4 g above 20 Hz in the horizontal direction),
- operating theatre (ISO): 0.13 mm/s between frequencies from 60 to 1000 Hz,

- analytical balance: 0.076 mm/s between frequencies from 45 to 1000 Hz,
- electronic microscope (Phillips): 0.025 mm/s between frequencies from 50 to 1000 Hz

  Amick (1997) and BS 5228-2 (2009) provide the following limits:

- optical microscope with magnification 400 times, microbalances, optical balances, proximity and projection aligners etc.: 0.050 mm/s at 8+ Hz
- optical microscope with magnification 1000 times, inspection and lithography equipment (including steppers) to 3 μm line width: 0.025 mm/s at 8+ Hz
- most lithography and inspection equipment (including electron microscopes) to 1 μm detail size: 0.0125 mm/s at 8+ Hz
- electron microscopes (TEM's and SEMs) and E-Beam systems: 0.006 mm/s at 8+ Hz
- long path laser based small target systems 0.003 mm/s at 8+ Hz

It is worth mentioning that footfall induced floor vibration velocity is in the range from 1.1 to 3.8 mm/s between frequencies from 5 to 10 Hz according to Dowding (2000). New (1986), instead, reports peak particle velocities between 0.02 and 0.5 mm/s from footfalls, 0.15 to 3.0 mm/s from foot stamping, 3 to 17 mm/s from door slamming and 5 to 20 mm/s from percussive drilling in buildings. Not only precise equipment but also other industrial machine manufacturers specify tolerable levels of vibration. For example,

- large compressor (MAN) foundation velocity ≤2.8 mm/s in operational condition and ≤6 mm/s in accidental case between frequencies from 25 to 190 Hz
- gas turbine (EGT) foundation velocity ≤2 mm/s and that a peak to peak amplitude of any part of the foundation is less than 50 μm at the operating frequency of 250 Hz.

The effect of foundations on amplification/attenuation of ground vibration is considered in Section 5.2.

1.2.3 Structures

Vibration can cause from cosmetic (superficial) damage of plaster on walls to serious structural damage. A rather comprehensive overview of codes and standards related to ground vibrations is provided by Skipp (1998), New (1986), Hiller and Crabb (2000), etc. The effects of structures on amplification/attenuation of ground vibration are considered in Section 5.3.

1.2.3.1 Examples of Guidelines in Standards Used Internationally

- German DIN 4150-3 (1999) specifies peak velocities of foundations by transient vibrations causing so called cosmetic damage (opening of cracks in plaster on

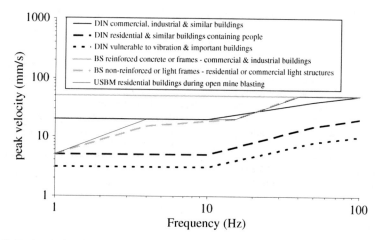

Fig. 1.2 Peak velocity of foundations/basements for appearance of cosmetic cracking in buildings due to transient vibration

walls, increase of existing cracks, and detachment of non-structural partitions from structural walls and columns) as shown in Fig. 1.2.

- British BS 7385-2 (1993) specifies peak velocities of building bases arising from transient vibrations causing cosmetic damage to buildings as shown in Fig. 1.2. For non-reinforced or light frames, at frequencies below 4 Hz, a maximum displacement of 0.6 mm (zero to peak) should not be exceeded.
- USBM (U.S. Bureau of Mines) RI 8507 (1980) specifies peak velocities causing visible damage to residential houses as a result of open mine blasting as shown in Fig. 1.2.

British standard BS 5228-2 (2009) recommends the threshold peak particle velocities for minor or cosmetic (i.e. non-structural) damage shown in Table 1.2.

1.2.4 Collapsible/Liquefiable/Thixotropic Soil

When soil with rather uniform grain sizes and consequently with large void spacing between grains is subjected to vibration its grains tend to move into adjacent voids because of the unstable structure of such soil.

- If collapsible soil is above ground water level then it can exhibit significant settlement causing subsidence of buildings with shallow foundations.
- If collapsible soil is under ground water level then its tendency to decrease its volume by its grains moving into adjacent voids leads to build up of excessive water pressure if water flow is prevented. Soil shear strength is proportional to the difference between total pressure due to soil weight and excess pore

Table 1.2 Threshold peak particle velocities in mm/s for minor or cosmetic damage according to BS 5228-2 (2009)

Vibration type	Reinforced or framed structures. Industrial and heavy commercial buildings	Not reinforced or light framed structures. Residential or light commercial buildings	Slender and potentially sensitive masonry walls	Propped or tied walls or mass gravity walls	Underground services (for elderly and dilapidated brickwork sewers to use 20–50% reduction)
Intermittent vibration	50	15@4 Hz 20@15 Hz, 50@40 Hz	10@the toe 40@the crest	50–100% Greater than for slender and potentially sensitive masonry walls	30
Continuous vibration	50% lower than the intermittent vibration limits		Reduced 1.5–2.5 the intermittent vibration limits		15

water pressure. If excess pore water pressure equals the total pressure then soil shear strength decreases to zero and it behaves like a heavy fluid i.e. liquefies (e.g. Seed, 1979). Liquefied soil can cause large distance flow failure of slopes, sinking/tilting of buildings, excessive lateral pressures on and failure/large deformation of retaining walls/pile groups adjacent to sloping ground. Even if soil does not liquefy, increase in excess pore water pressure can lead to softening of medium dense soil and consequently to large deformations.

- Thixotropy is defined as an isothermal, reversible, time-dependent process which occurs under constant volume when a material softens instantly, as a result of disturbance including shaking, and then gradually returns to its original strength when allowed to rest. It should be noted that thixotropy occurs under constant soil volume unlike liquefaction, which requires decrease in soil volume. Clay with natural water content close to the water content corresponding to its liquid state is known to be subjected to almost complete shear strength loss when disturbed. Long distance flow type failures in so called quick clay are well known (e.g. Ter-Stepanian, 2000). Seed and Chan (1959) demonstrated that thixotropic strength regain is also possible for soil with water content at or near the limit of its plasticity. More information on formation of quick clay in Sweden, for example, is provided by Rankka et al. (2004).

Because of a large number of factors affecting behaviour of sensitive soil, there are no well establish criteria for limited ground vibration to prevent collapse of such soil. Two case histories of soil liquefaction caused by machine vibrations considered by Olson (2001) are described for example.

1.2.4.1 Example of Failure of an Embankment in Sweden

Ekstrom and Olofsson (1985), Konrad and Watts (1995) and Yashima et al. (1997) described and consider the failure of a road embankment during pavement repair near Asele in northern Sweden on 4 October 1983. The embankment is located in a partially impounded reservoir for a new hydro-electric power station. The slide was triggered by a tractor-towed 3.3 ton vibratory roller, which slid into the reservoir along with the embankment killing the operator. Flow slides in submerged loose sandy soil placed hydraulically or loaded by fill on top of it were caused not only by machinery but also rapid construction (e.g. Olson, 2001).

According to Ekstrom and Olofsson (1985), the slide at Asele occurred very suddenly, lasting about 10 s, on the first pass of the vibratory roller. Ekstrom and Olofson (1985) suggest that freezing of the fill placed using a 'wet fill' method during winter months prevented proper compaction resulting in a loose soil structure. Grain size distribution of till used for embankment fill indicates maximum grain size 20 mm, 89–98% grains smaller than 10 mm, 60–74% grains smaller than 1 mm and 20–37% grains smaller than 0.06 mm. Konrad and Watts (1995) indicated that the embankment fill had an average standardized blow count number (described in Section 7.2.1.1) from standard penetration tests (e.g. ASTM D1586; Eurocode 7–part 2) of approximately 6–8. Yashima et al. (1997) stated that the 3.3 ton vibrating roller was sufficient to cause large excess pore water pressure increase in the fill and trigger flow failure. Details of the vibratory roller used are not known.

1.2.4.2 Example of Failure of an Embankment in Michigan, USA

Hryciw et al. (1990) described a unique liquefaction flow failure of a road embankment made of hydraulically placed sand fill on 24 July 1987. The failure was triggered by six 196 kN (22 ton) trucks conducting a seismic reflection survey of which three fell into the Lake Ackerman. The sudden failure of the embankment created a 4.5 m high wave that crossed the 122 m long lake and destroyed a boat dock, which remains were strewn about 9 m into the woods.

The fill material was medium to fine sand, with grain sizes mainly in the range from 1 to 0.2 mm, end-dumped into the lake and moderately compacted above the lake water level. Standardized blow count number (described in Section 7.2.1.1) of the standard penetration tests ranged from 1 to 7 below the lake water level and from 3 to 11 above the lake water level.

1.3 Frequent Sources of Ground Vibration

1.3.1 Construction/Demolition Activities

Construction/demolition activities causing problems in vicinity of structures and soil sensitive to vibration are:

- pile driving
- soil shallow compaction
- demolition of structures
- rock excavation by explosives
- soil deep compaction by explosives

1.3.1.1 Pile Driving

Two main types of pile driving hammers, impact and vibration based, are shown in Fig. 1.3 together with ground wave fronts (shaded zones) of transmitted vibration energy away from piles.

The use of impact hammers causes emission of transient waves along pile shaft and from pile toe each time pile driving hammer hits the cushion on pile top. The pile driving energy transmitted from a pile shaft into surrounding ground as vertical transversal waves is proportional to the product of the shaft friction force and pile vertical displacement, which both vary along depth; the shaft friction force increases with depth and pile displacement decreasing with depth because of pile elastic shortening. The stress pulse that is generated by hammer impact at pile top propagates along a steel pile at about 5.1 km/s and along a concrete pile at about 2.9 km/s. For example, the stress pulse travels 0.004 s along 20 m long pile if it is made of steel and 0.007 s if the pile is made of concrete. As ground waves propagate at about one order of magnitude (ten times) slower than the speed of stress pulse along a

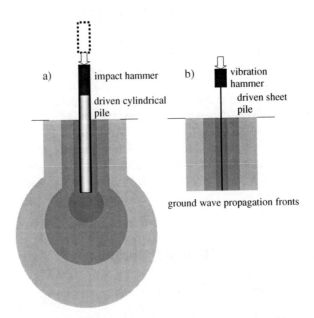

Fig. 1.3 Schematic pile driving using impact (**a**) and vibration (**b**) based hammers with cross sections of fronts of propagating waves in homogeneous ground

pile, it will be assumed in a simplified approach that the driving energy is uniformly distributed along pile shaft. Jaksa et al. (2002), for example, measured acceleration time histories of ground vibration during installation of enlarged base driven cast in situ pile (Frankie type) and obtained the distribution of peak particle velocities along depth and around the pile similar to the distributions of wave fronts shown in Fig. 1.3a.

Pile driving energy emitted from a pile toe is proportional to the product of force and displacement at pile toe. Both longitudinal and transversal ground waves are generated at pile toe because of the Poisson's effect in a continuum resulting in soil compression accompaniment by shearing. Attewell and Farmer (1973) observed that the pile driving involved generation of compression waves that propagate from the area of the pile toe and expand outwards over a spherical front. In a simplified approach, pile toe will be considered as a point source of energy emition into surrounding ground, as shown in Fig. 1.3a.

The use of vibration hammers causes continuous vibration of ground and emission of the vertical shear waves from the surface of sheet piles, Fig. 1.3b. One or more pairs of horizontally opposed contra-rotating eccentric weights are used by vibration hammers to cause ground vibration in order to reduce the friction between the pile and soil so that combined weight of the pile and hammer cause the pile to penetrate into ground. Operating frequencies of vibration hammers are between 25 and 50 Hz (e.g. Hiller and Crabb, 2000) i.e. between 0 and 1400 Hz (e.g. Dowding, 2000). The energy emitted from sheet pile surface is proportional to the product of the shaft friction force and pile vertical displacement, which both vary with depth.

1.3.1.2 Soil Shallow Compaction

The following main types of shallow soil compaction can cause significant ground vibrations:

- compaction of fill layers by vibrating rollers
- dropping of heavy weights (dynamic compaction)
- vibro float (pendulum like probe oscillation) and vibro rod
- compaction piles

The use of compaction piles is equivalent to pile driving using impact hammers and the use of vibro rod is equivalent to pile driving using vibrating hammers. Both compaction of fill layers by vibrating rollers and dynamic compaction by dropping heavy weight cause mainly near surface propagating ground waves shown in Fig. 1.4a. Vibro floatation uses pendulum like probe to penetrate and compact soil in horizontal direction and therefore generates mainly horizontal axial and transversal waves, which propagation fronts are shown in Fig. 1.4b.

Heavy weight tamping causes transient ground vibration with the impact energy equal to the product of dropped weight and height, which are easy to determine. Vibration rollers cause continuous ground vibration (in the frequency range between 0 and 53 Hz, e.g. Hiller and Crabb, 2000) with the source energy equal to the

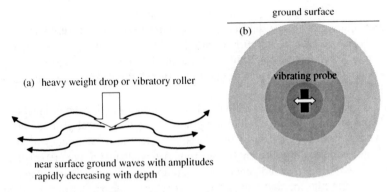

Fig. 1.4 (**a**) Sketch of near surface ground waves induced by load acting on the surface, (**b**) cross sections of fronts of propagating waves in homogeneous ground from horizontally vibrating probe used for vibrofloatation

dynamic weight and roller vibration amplitude, which are specified by the manufacturer and can be verified when required. The energy (product of applied force and soil displacement) at the location of vibrating probe (with operating frequencies usually between 30 and 50 Hz) varies with change of soil density and is rather difficult to assess. Instead, the rated energy specified by the manufacturer of the probe multiplied by an efficiency coefficient can be used for simplified analyses of propagation of continuous ground waves from such approximated point source.

1.3.1.3 Demolition of Structures

Ground vibration caused by demolition of existing structures arises mainly from the use of:

- pneumatic hammers mounted on vehicles
- explosive to bring structures down

Ground vibration arising from the use of pneumatic hammer to demolish a structure depends on the vibration of a whole structure and not only on its part being demolished and therefore is rather difficult to predict. Despite great weight of vibrating structure, the vibration amplitudes are rather small and therefore the energy emitted at the source should be small. The noise created by use of pneumatic hammer could create greater problem than vibration and, therefore, influence a decrease in applied dynamic forces.

Demolishing of structures using explosive charges is rare but happens occasionally. Eldred and Skipp (1998), for example, summarised several case histories of demolition of several cooling towers and large chimneys and showed that the ground vibration caused by explosive demolition is a series of connected transient/continuous events arising from fall of larger structural blocks/elements and with total duration of about 10 s. The greatest recorded peak velocity in one case was 952 mm/s at frequency of 4.5 Hz and at the source to instrument distance of 12.6 m. The energy released at the source is proportional to the product between the weight of larger structural blocks/elements and their height above ground and is transmitted mostly by near surface ground waves sketched in Fig. 1.4a.

1.3.1.4 Blasting in Construction and Mining Industries

Explosives are used in both construction and mining industry for digging tunnels in rock (e.g. Hoek and Brown, 1980; Gregory, 1984) and for cutting slopes of trenches in rock (e.g. Hoek and Bray, 1981; Gregory, 1984). The transient ground waves caused by the use of explosives are dependent on the energy of explosions of individual charges, which is specified by the manufacturers of explosives. In simplified approach, the explosion at the face of a tunnel can be considered coming from a point source as shown in Fig. 1.4b for vibratory probe. The explosions used to cut rock slopes are considered in the simplified approach as a planar energy source as shown in Fig. 1.5.

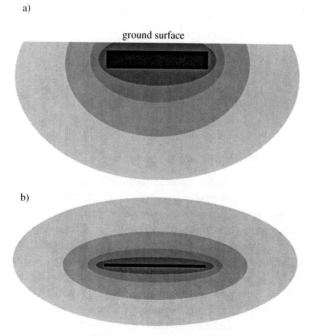

Fig. 1.5 (a) Vertical cross section, and (b) layout of the fronts of propagating waves in homogeneous ground from a planar energy source (*black colour*) arising from cutting of rock slopes by explosive charges

Fig. 1.6 Sketch of propagation of the fronts of waves in homogeneous ground from a prismatic source caused by soil deep compaction by explosives

ground surface

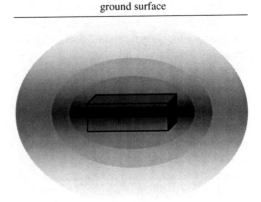

1.3.1.5 Soil Deep Compaction by Explosives

Gohl et al. (2000) and Towhata (2007) provide more information on this technique, which causes transient ground vibrations. The energy released by explosives is specified by the manufacturers. In the simplified analyses, prismatic volume of compacted soil is considered as a source shown in Fig. 1.6.

1.3.2 Traffic

Two types of traffic vibration sources are considered

- trains
- road vehicles

1.3.2.1 Train Induced Vibrations

Holm et al. (2002), for example, reported on mitigation of track and ground vibrations by high speed trains at Ledsgard, Sweden. The site contains pocket of very soft organic soil (gyttja) up to 3 m thick below a dry crust. The gyttja is underlain by soft clay and the depth to bedrock is more than 60 m. The results of measurements of the track settlement and accelerations at different train speed indicated that the dynamic amplification is insignificant up to about 140 km/h and that a significant increase occurs at about 180 km/h. These values are appropriate at Ledsgard and may be different for other ground conditions. Holm et al. (2002) also stated that focusing of wave energy can occurs when the bedrock is concave and acts like wave reflector as well as that horizontal soil layers may refract waves upwards so that they meet at the surface at close distance. Holm et al. (2002) suggested that topographic effects that amplify ground vibration can be found on hilltops and in soil at the foot of outcropping rock. An increased level of vibration can occur at the rim of a river

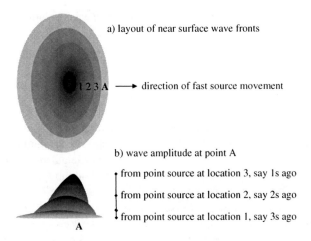

a) layout of near surface wave fronts

1 2 3 A ⟶ direction of fast source movement

b) wave amplitude at point A

from point source at location 3, say 1s ago

from point source at location 2, say 2s ago

from point source at location 1, say 3s ago

A

Fig. 1.7 Sketch of the superposition effect from fast point source movement in the direction of movement on ground wave amplitudes

valley or gully. Holm et al. (2002) discussed possible effect on ground vibration of train speed relative to the speed of propagation of body and surface ground waves. A sketch shown in Fig. 1.7 illustrates the effect of high speed motion of a point source on the amplitude of ground wave, which essentially is superposition of amplitudes of radiated waves at different source locations in time.

Bahrekazemi (2004), for example, provided the state of the art review of ground vibration induced by trains. He referred to Dawn and Stanworth (1979) who stated that if trains were to travel faster than the propagation velocity of the ground vibration, the shock wave which is formed in the ground would seriously affect the nearby buildings. It was suggested by Dawn and Stanworth (1979) that the excitation of ground vibration, especially at low frequencies, depends on the total vehicle mass, not just the un-sprung mass of the wheel set. This was evidenced by a large measured difference between loaded and unloaded trains.

In the simplified approach, the energy at the source causing vibration is considered proportional to the product of the force acting per train axle and ground surface settlement caused by such force. The force from train weight is increased by inertia force for high speed trains. The inertia force is a product of the acting mass per axle (the force from train weight divided by the gravitational acceleration) and ground surface acceleration. The ground vibration from train movement is mainly transmitted as surface waves, sketched in Fig. 1.4a. High speed trains also effect frequency/wave length of ground wave propagation according to Doppler effect arising when the source, observer or propagation medium moves. For example, http://en.wikipedia.org/wiki/Doppler_effect states that: *The receiver frequency is increased (compared to the emitted frequency) during the approach, it is identical at the instant of passing by, and it is decreased during the recession.*

Underground trains (metros) also induce vibrations, which can cause annoyance to the residents of the buildings placed above them or disturbance to precise processes such as operating theatres, in which case special isolation devices have to be used to minimise the vibrations.

1.3.2.2 Vehicle Induced Vibrations

Watts (1987), for example, reported on vibration measurements at a number of houses from a sample of 1600 that were identified from an earlier survey of vibration nuisance at 50 residential sites. Heavy goods vehicles and buses produce most noticeable vibration. The vibration levels from such vehicles tend to increase with vehicle speed as well as with increase in maximum height or depth of the road surface irregularity, particularly within 5 m from buildings and the irregularity depth/height greater than approximately 20 mm. Watts (1987) also stated that *At none of the measurement sites did passing heavy vehicle produce peak levels of vibration near the façade or on the ground floor which exceeded one of the lowest thresholds for minor damage that has been proposed for this type of building.* However, for non-maintained roads with deep holes, significant vibrations may be generated by the impact of wheels of heavy vehicles.

In the simplified approach, the energy released at the source of transient and near surface waves shown in Fig. 1.4a is considered proportional to the product of wheel/axle load and depth of road hole i.e. height of a bump (laying policeman) used for traffic calming purposes, Fig. 1.8.

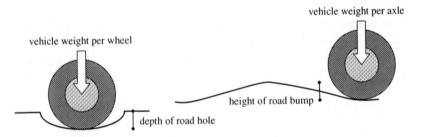

Fig. 1.8 Sketches of the source of ground vibration sources caused by traffic movement

1.3.3 Machinery

Transient vibration is caused by industrial hammers while continuous vibration by eccentric parts of rotating machinery such as gas turbines and pumps and unbalanced parts of reciprocating (translational) motion of pistons of compressors if they are not base isolated. Manufacturers of machinery specify both acceptable vibration parameters (Section 1.2.2) and dynamic loads (unbalanced masses) at different frequencies. Several standards exist for design of foundations for machinery (e.g. CP 2012, 1974; DIN 4024-1, 1988; DIN 4024-2, 1991). Shallow machine foundations generate mainly near surface waves as sketched in Fig. 1.4a while deep (mainly piled) foundation induce both body and surface waves similar to the waves generated by pile driving (Fig. 1.3a) but of continuous rather than transient type.

1.3.3.1 Examples of Dynamic Loads From Machinery

Dynamic unbalanced forces occurring at accidental load from a large compressor (MAN) are given in Table 1.3.

Dynamic loads acting during operation of a gas turbine (EGT) are given in Table 1.4.

Table 1.3 Accidental dynamic forces for a large compressor

Component	Frequency (Hz)	Unbalanced force (N)
Driver rotor	25	41650
Wheel shaft	25	27160
Pinion shaft	194	9430
Compressor rotor	191	16835

Table 1.4 Example loads generated by operation of a gas turbine

Description	No 1	No 2	No 3	No 4	No 5	No 6	No 7
Unbalanced force amplitude (N)	1148	3265	2980	5004	7073	5004	7073
Moment amplitude (Nm) from the horizontal and vertical forces around the centre of gravity	1251	3559	6943	5454	16480	5454	16480
Frequency (Hz)	250	250	250	238	238	238	238

1.4 Vibration Propagation Media Effects

Two vibration propagation media effects will be considered:

- ground
- foundation

Vibration propagation through structures is beyond the scope of this volume.

1.4.1 Ground Amplification and Attenuation of Wave Amplitudes

1.4.1.1 Wave Amplitudes Amplification

Amplification of amplitudes of ground waves could be caused by several factors such as:

- impedance contrast (Section 2.4.1) when refracted waves (Section 2.4.2) propagate from layers with higher velocity into layers with lower velocity,
- superposition of wave amplitudes because of wave reflection (Section 2.4.3) from bedrock, soil layers and topographic features and their focusing at a location,
- resonance between ground natural vibration frequency and the frequency of wave generator. Famous anecdotic cases of resonance aided amplification of structural and ground vibration are the 'artificial earthquakes' in Chicago caused by Tesla's resonance machine of minimal weight and size (e.g. http://www.excludedmiddle.com/earthquake.htm)

A single degree of freedom oscillator (SDOFO) subjected to harmonic load is used to visualize the effect of resonance on amplification of amplitudes of vibration, Fig. 1.9.

The ratio between the peak output acceleration a_o and the corresponding peak input acceleration a_i of a SDOFO is shown in Fig. 1.10.

The amplification factor for harmonic motion of a SDOFO (e.g. Clough and Penzien, 1993) is:

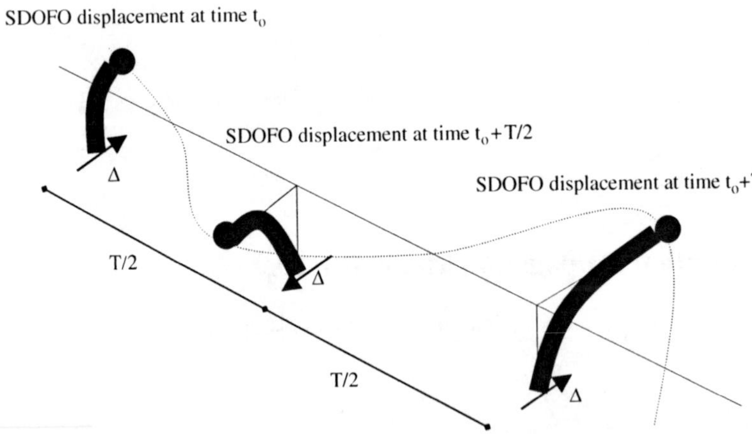

Fig. 1.9 Addition of relatively small displacement amplitude Δ by a vibrator (*arrow*) in resonance with SDOFO (cantilever beam with lumped mass at the top) every half a cycle (with period T) during a harmonic vibration causes an exponential increase in SDOFO displacement amplitudes in time

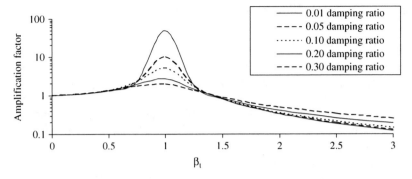

Fig. 1.10 Amplification factor between the output and input peak accelerations of a SDOFO

$$\frac{a_o}{a_i} = \sqrt{\frac{1 + (2\beta_t\xi)^2}{(1 - \beta_t^2)^2 + (2\beta_t\xi)^2}} \quad , \tag{1.1}$$

where $\beta_t = f_d f_o^{-1}$ is the tuning ratio, f_d is the frequency of an input motion i.e. vibrator, f_o is the frequency of the output motion i.e. ground, ξ is the damping ratio (Section 2.4.7), a_o is the acceleration amplitude of SDOFO, a_i is the acceleration amplitude of input motion (ground). Damping ratio is the ratio between actual and critical damping, which prevents ground vibration.

1.4.1.2 Wave Amplitude Attenuation

Attenuation of wave amplitudes by ground is the result of radiation (Section 2.4.6) and material damping (Section 2.4.7).

Waves tend to spread from the source for the propagation medium to achieve the state of a minimum energy (e.g. http://en.wikipedia.org/wiki/Principle_of_minimum_energy). As waves spread, the principle of energy conservation (e.g. http://en.wikipedia.org/wiki/Conservation_of_energy) states that their amplitudes must decrease because the wave energy is proportional to the square of wave amplitude (e.g. Bormann, 2002) and the total energy along the wave front must remain the same as the front spreads.

Material damping causes decrease of wave amplitudes during wave propagation through ground as a result of the friction between ground particles. Material damping is usually small, less than one percent in rock or a few percent in soft/loose soil at small strain. At large strain (generated in close vicinity of pile driving, soil compaction etc.) soil damping could increase up to about 30% of the critical damping.

1.4.2 Foundation Kinematic and Inertia Interactions

1.4.2.1 Kinematic Interaction

Kinematic interaction (Section 5.2) happens when the stiffness of a foundation (and the structure above it) is greater than the stiffness of underlying ground so that the foundation (and the structure above it) cannot experience the same deformation (wave amplitudes) as the underlying ground beneath them. Stiff foundation (and structure) tend to average uneven ground motion amplitudes beneath them and as a result suffer increase in internal stresses, Fig. 1.11.

Displacement amplitudes Δ_w are considered in Section 2.3. Wave length $L_w = v\,T_w$, where v is wave velocity and T_w is wave period, which depends on the fundamental period of ground vibration (Section 4.4) and the period of vibration of the source.

The shear stress increment $d\tau = G\,\Delta_w\,L_f^{-1}$, where G is shear modulus of foundation (and structure), Δ_w is amplitude of ground displacement at the corner of foundation (and structure), L_f is foundation length. Practical problem is to determine not only the displacement amplitude Δ_w but also the initial minor and major axial stresses before ground motion particularly for complicated geometries with window and door openings in multi-storey buildings, in which case a numerical

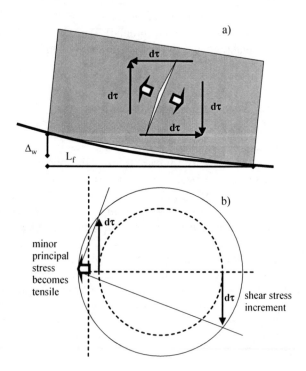

Fig. 1.11 (a) Stiff foundation (structure) subjected to wave with amplitude Δ_w causing shear stress increment $d\tau$ (*black arrow*), tensile minor principal stress (*white arrow*) and a diagonal crack when the stress exceeds the strength in tension, (b) Mohr circle

method may need to be used. Mirror image is obtained when the opposite side of foundation (and structure) has been lifted by ground wave amplitude.

The pattern of a foundation (structure) movement/crack shown in Fig. 1.11 can also be caused by, for example, uplifting of its one corner due to swelling of underlying clay when it is saturated by a broken gutter or water supply pipe or by subsidence of its other corner resting on a collapsible soil when saturated by water or shrinkable soil due to water extraction by tree roots. Because of similarity of crack patterns in steady and cyclic conditions, it is important to know the state of a foundation (structure) before arrival of ground waves from a vibration source.

Not only stiff foundation (and structures above them) may develop diagonal cracks on ground wave passage beneath them but also flexible foundations (and structures above them) may develop vertical top and bottom cracks if the principal horizontal axial tensile stress exceeds the material strength, as shown in Fig. 1.12.

The maximum incremental tensile horizontal stress σ at the underside of a homogeneous foundation (structure) is according to the linear engineering beam theory (e.g. http://en.wikipedia.org/wiki/Euler%E2%80%93Bernoulli_beam_equation). The theory is not applicable to thick beams.

$$\sigma_t = \frac{E \cdot {}^{H_f}/2}{\dfrac{L_f^2}{8 \cdot \Delta_w} + \dfrac{\Delta_w}{2}} \quad , \tag{1.2}$$

where E is Young modulus of foundation (structure), H_f is foundation (structure) height, Δ_w is amplitude of ground displacement at the middle of foundation (structure), L_f is foundation length. When the resultant tensile horizontal stress exceeds the tensile strength, near vertical crack(s) occur. For more complicated shapes of foundations (structures with window and door openings) a numerical method may need to be used for the calculation of axial tensile stress σ_t. Also when the displacement amplitude Δ_w is oriented upwards instead of downwards as in Fig. 1.12, near vertical crack(s) occur(s) at the top of foundation (structure) instead at the bottom.

The pattern of sagging/hogging and cracking shown in Fig. 1.12 due to wave propagation can also be caused by other non-dynamic reasons like moisture content change beneath foundation (structure) in collapsible/shrinkable/swelling soil,

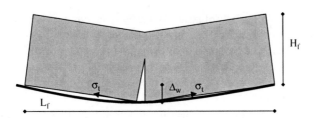

Fig. 1.12 Near vertical crack in a flexural foundation (structure) due to ground wave amplitude Δ_w

seasonal freezing/thawing in the case of shallow foundations, insufficient bearing capacity of foundations (if any), etc. Because of similarity of crack patterns in steady and cyclic conditions, it is important to know the state of a foundation (structure) before arrival of ground waves from a vibration source.

1.4.2.2 Inertial Interaction

Inertial interaction (Section 5.2) is caused by foundation (and structure) mass inertia effects. According to Huygens-Fresnel principle (e.g. http://en. wikipedia.org/wiki/Huygens-Fresnel_principle) each point subjected to vibration effect becomes a source of vibration itself. Because foundations (and structures) have mass and stiffness different from ground mass and stiffness, foundations (and structures) will vibrate with different frequency and radiate waves of different amplitudes when subjected to ground vibration. As a result of the foundation (and structure) inertia effect, the superposition of incoming waves from a source and out-going waves from foundation (and structure) will cause interference and result in increased or decreased wave amplitudes. In order to avoid increase in amplitudes of resulting waves it is important to avoid resonance effect i.e. the same frequency of vibration of foundation (and structure) and ground vibration.

As a rule of thumb, the frequency of vibration of steel and concrete moment resisting frames in buildings is proportional to the ratio between five and the number of storeys while for concrete shear walls in building to the ratio between 10 and the number of storeys (e.g. Shakal et al., 1996). Eurocode 8–Part 1 suggests that the frequency of vibration of buildings is $13.33H_b^{-3/4}$, where H_b is building height. For non-building structures, their frequency of free vibration can be estimated based on the expression given in Section 5.3. Free vibration frequency of soil layers can be assessed based on the expression given in Section 2.2.

1.5 Summary

This introductory chapter describes frequent sources of ground vibrations, basic propagation media effects and limits/sensitivities of vibration recipients.

- Of all considered recipients of ground vibrations, electronic microscopes in build-ings are the most sensitive ones allowing the peak floor velocity of up to about 0.025 mm/s, followed by humans (in operating theatres) and analytical balances at about 1 mm/s, vulnerable buildings and industrial machines allow up to about 2–5 mm/s and finally commercial and modern industrial building up to 50 mm/s of the peak velocity.
- Continuous vibration imposes lower limited peak velocities than transient vibra-tion, roughly about 50% lower, because of fatigue effects on materials during repetitive loading.

- Of frequent sources of ground vibrations, pile driving and soil compaction (including blasting) induces largest deformations and vibrations particularly in vicinity of the piles and compaction locations, followed by traffic and various industrial machines. Simplified analyses of wave propagation from idealized point, linear, planar and prismatic sources are performed using formulas and spreadsheets provided in the Appendices of this volume.
- Ground and foundations (with structures above them) are capable of both amplification and attenuation of amplitudes of incoming waves depending on a number of factors, which are described in more details in Section 1.4. It is important to avoid amplification of amplitudes of ground waves by resonance effect i.e. similar frequencies of vibrations of the source, the adjacent ground and the receiver i.e. foundations (including structures).

Chapter 2
Ground Waves Propagation

2.1 Introduction

Ground waves transmit energy from vibration sources, which are described in Section 1.3. The transmitting of vibration energy occurs because ground tends to reach the state of a minimum energy when disturbed by vibration (e.g. http://en.wikipedia.org/wiki/Principle_of_minimum_energy). Ground disturbance by a vibration source causes occurrence of stress waves, which transmit the source energy in the form of energy flux.

The total energy-flux density E_{flux} per unit time (i.e. wave power) in direction of wave propagation through wave front area dS is in the case of an isotropic stress-strain relationship in a non-dispersive (closed) system (e.g. Bormann, 2002)

$$E_{flux} = \frac{1}{2} \cdot v \cdot \rho \cdot \Delta_w^2 \cdot \omega^2 \cdot dS, \qquad (2.1)$$

where v is ground wave propagation velocity, ρ is ground density, Δ_w ground displacement amplitude, ω circular frequency of ground vibration $= 2\pi f$.

The longitudinal velocity v_l of axial stress waves is different from ground particle velocity $\partial \Delta_{w,l}(\partial t)^{-1}$ in direction of wave propagation. The particle velocity $\partial \Delta_{w,l}(\partial t)^{-1}$ in the longitudinal direction can be determined from the following relationships (e.g. Kramer, 1996) for an infinitely long linear elastic rod considered for simplicity

- longitudinal strain-displacement relationship $\partial \Delta_{w,l} = \varepsilon_l \partial l$
- axial linear elastic stress-strain relationship $\varepsilon_l = \sigma_l E^{-1}$
- longitudinal wave propagation velocity $\partial l = v_l \partial t$

as

$$\frac{\partial \Delta_{w,l}}{\partial t} = \frac{\varepsilon_l \cdot \partial l}{\partial t} = \frac{\sigma_l}{E} \cdot v_l = \frac{\sigma_l}{\rho \cdot v_l^2} \cdot v_l = \frac{\sigma_l}{\rho \cdot v_l}, \qquad (2.2)$$

where symbol ∂ denotes partial derivative, ε_l is axial strain i.e. the ratio between longitudinal displacement and the length over which it has been achieved, σ_l is

M. Srbulov, *Ground Vibration Engineering*, Geotechnical, Geological, and Earthquake Engineering 12, DOI 10.1007/978-90-481-9082-9_2, © Springer Science+Business Media B.V. 2010

axial stress i.e. the ratio between axial force and the area on which it is acting, ρ is unit density, v_l is velocity of longitudinal waves, t is time.

In practice, it is more usual to measure particle velocity $\partial \Delta_{w;l}(\partial t)^{-1}$ instead of the axial stress σ_l in direction of wave propagation. When particle acceleration $\partial^2 \Delta_{w;l}(\partial t^2)^{-1}$ in direction of wave propagation is measured then $\partial^2 \Delta_{w;l}(\partial t^2)^{-1} = \rho^{-1}\partial\sigma_l(\partial l)^{-1}$ (e.g. Kramer, 1996) for an infinitely long linear elastic rod considered for simplicity. Following Timoshenko and Goodier (1970), the particle velocity $\partial \Delta_{w;l}(\partial t)^{-1}$ of a transversal wave is

$$\frac{\partial \Delta_{w;t}}{\partial t} = \frac{\tau}{\rho \cdot v_t}, \qquad (2.3)$$

where τ is shear stresses i.e. the ratio between transversal force and the area on which it is acting, v_t is transversal wave velocity, ρ is unit ground density, t is time.

The objective of this chapter is to provide descriptions with comments on types and amplitudes of ground waves, of ground wave path and other influential factor affects on ground wave propagation.

2.2 Main Wave Parameters

Waves are described by their maximum amplitude and spectra of amplitudes, frequency and duration of vibration.

- Amplitudes of ground waves vary both in space and time. For simplified analysis and in practice, the maximum vertical component of ground displacement/velocity/acceleration is frequently considered although this component may not be the largest one. Other choices for considerations are the largest of the three mutually perpendicular components, the maximum resultant value, which is the vector summation of the three components. Sometimes, the vector sum of the maximum of each component regardless of the time of their occurrence and the root-mean-square (r.m.s.) value, which is the square root of the average of sum of squares of componential values, are considered.
- Frequency is an important parameter of ground vibration not only because the sensitivities of recipients (Section 1.2) depend on vibration frequency but also because ground displacement, velocity and acceleration are related among themselves depending on frequency (or period) of vibration. For (an equivalent) harmonic vibration, which amplitude variation in time and space is described by a sine function, the ground displacement amplitude is proportional to the ratio between ground velocity amplitude and the circular frequency $\omega = 2\pi f$ of the vibration and ground velocity amplitude is proportional to the ratio between ground acceleration amplitude and the circular frequency ω of the vibration.
- Frequency of ground vibration depends on both the frequency of vibration source and the frequency of free vibration of the layers. The circular frequency $\omega_n = 2\pi f_n$ and period $T_n = f_n^{-1}$ of the nth mode of free vibration of an infinite

Fig. 2.1 Normalised
horizontal displacements at a
time instant for the first three
modes of vibration of a
horizontal layer

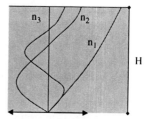

layer, with constant soil properties over an interval of shear strain is (e.g. Srbulov,
2008)

$$\omega_n = \frac{2 \cdot \pi \cdot (2 \cdot n_m - 1)}{4 \cdot H} \cdot \sqrt{\frac{G}{\rho}}$$
$$T_n = \frac{2 \cdot \pi}{\omega_n},$$

(2.4)

where n_m is vibration mode number, H is soil layer thickness, G is shear modulus,
ρ is unit soil density. The normalised horizontal displacements for the first three
vibration modes in the horizontal direction of a horizontal layer are shown in
Fig. 2.1.

The fundamental resonant frequency of the top soil layers and an estimation of
local site amplification of ground motion can be determined from micro-tremor
measurements as described in Section 3.2.1.1.

- Duration of vibration affects wave impact on recipients. Continuous waves have
more negative impact than transient waves because of the fatigue effect on both
people and materials. Also, continuous vibration can cause resonance effects as
indicated in Section 1.4.1.1.

2.3 Types and Amplitudes of Ground Waves

2.3.1 Body Waves

Body waves (i.e. waves at depth) originate from deep vibration sources such as
described in Sections 1.3.1.1–1.3.1.5, for example. Two types of body waves exist –
longitudinal and transversal.

- Longitudinal waves propagate in direction of the particle movement and are
caused by axial stresses originating from pile toes (Fig. 1.3a), vibrating probe
(Fig. 1.4b), cutting of rock by explosive charges (Fig. 1.5) and soil compaction
by explosives (Fig. 1.6) for example.
- Transversal waves propagate perpendicular to direction of the particle movement
and are caused by shear stresses present along pile shafts (Fig. 1.3) and because of
Poisson's effect at: pile toes, around vibrating probes, cutting of rock by explosive
charges and soil compaction by explosives.

Longitudinal waves travel faster than the transversal waves. The ratio between velocities of the longitudinal v_l and transversal v_t waves in an elastic solid is (e.g. Kramer, 1996)

$$\frac{v_l}{v_t} = \sqrt{\frac{2 - 2 \cdot v}{1 - 2 \cdot v}},\tag{2.5}$$

where v is Poisson's ratio. For Poisson's ratio of 0.5, which is characteristic for fully saturated soft to firm clay in undrained condition (when there is no time for excess pore water pressure caused by wave propagation to dissipate), the velocity v_l would be infinite according to Equation (2.5), which is not realistic because Equation (2.5) is applicable to elastic solid with Poisson's ratio less than 0.5.

Propagation of body waves is a complex process due to existence of six different componential stresses at any location. For example, the equation of motion in the longitudinal direction in a three-dimensional elastic solid is (e.g. Kramer, 1996)

$$\rho \cdot \frac{\partial^2 \Delta_l}{\partial t^2} = \frac{\partial \sigma_l}{\partial l} + \frac{\partial \tau_{lt}}{\partial t} + \frac{\partial \tau_{lv}}{\partial v},\tag{2.6}$$

where ρ is ground density, $\partial^2 \Delta_l (\partial t^2)^{-1}$ is particle acceleration in the longitudinal direction, σ_l is axial stress in the longitudinal direction, τ_{lt} is shear stress in the plane perpendicular to the longitudinal direction, τ_{lv} is shear stress in the plane parallel to the longitudinal direction of ground wave propagation. Similar expression exists for the transversal t and other perpendicular direction to the longitudinal direction of ground wave propagation. Simplified one-dimensional analyses, which can not consider the shear stresses τ_{lt} and τ_{lv}, usually yield underestimated values of particle acceleration (e.g. Srbulov, 2008).

In an idealized homogeneous isotropic ground, the fronts of body waves are of spherical shape so that the ratio between the amplitude Δ_{wr} of ground displacement/velocity/acceleration at a distance r from an idealised point source location and the amplitude Δ_{wo} of ground displacement/velocity/acceleration at the source is proportional to the square root of the ratio of ground wave energy E_r per unit area at the distance r_s and the energy E_o at the source (e.g. Srbulov, 2008)

$$\frac{\Delta_{wr}}{\Delta_{wo}} \approx \sqrt{\frac{E_r}{E_o}} \approx \sqrt{\frac{1}{4 \cdot r_s^2 \cdot \pi \cdot e^{k \cdot r_s}}}\tag{2.7}$$

The expression $e^{\kappa r}$ describes wave energy loss due to its transformation into heat as a result of particle friction caused by wave propagation. The exponential function of ground motion amplitude on material damping has been established from laboratory test data. Typical values of attenuation coefficient κ range from a part of percent in rock to a few percent in soft/loose soil at small strain. Section 2.4.7 contains expression for calculation of the coefficient κ. Section 2.4 contains description of a number of factors affecting wave propagation in real ground, which is usually layered, anisotropic etc.

Ambraseys and Hendron (1968) considered the vibration from blasting using the Buckingham π theorem of dimensionless analysis (e.g. Langhaar, 1951) and found that the peak particle velocity $\partial \Delta_{wr}(\partial t)^{-1}$ is inversely proportional to the site to source distance r_s on cube power. Hence,

$$\frac{\partial \Delta_{wr}}{\partial t} \approx \sqrt{\frac{2}{\rho} \cdot \frac{E_o}{4/3 \cdot r_s^3 \cdot \pi \cdot e^{k \cdot r_s}}}, \tag{2.8}$$

where ρ is ground unit density and $4/3 r_s^3 \pi$ is the volume of ground between the source and the site. For distances from a vibration source larger than a half of wave length, consideration of half wave length $\lambda/2$ provides correct dimensional results i.e.

$$\frac{\partial \Delta_{wr}}{\partial t} \approx \sqrt{\frac{2}{\rho} \cdot \frac{E_o}{4 \cdot r_s^2 \cdot \pi \cdot \lambda/2 \cdot e^{k \cdot r_s}}} \tag{2.9}$$

In the idealized case, E_{flux} per unit area can be calculated as

$$E_{flux} = \frac{E_o}{4 \cdot r_s^2 \cdot \pi \cdot e^{k \cdot r_s}} \cdot \frac{v}{r_s}, \tag{2.10}$$

where v is ground wave propagation velocity.

For vibration sources of linear, planar and prismatic shape, the ground amplitudes ratio can be calculated using simple Gauss integration scheme (e.g. Zienkiewich and Taylor, 1991; Potts and Zdravkovic, 1999) for the characteristic points, which locations are shown in Fig. 2.2, instead of analytical or numerical integration over the source extent. For a linear vibration source that radiates energy uniformly, from Equation (2.8) and Fig. 2.2a with $0.5E_o$ at each point:

$$\frac{\Delta_{wr}}{\Delta t} = \sqrt{\frac{3 \cdot E_o}{4 \cdot \rho \cdot \pi} \cdot \sum_{i=1}^{2} \frac{1}{r_i^3 \cdot e^{k \cdot r_i}}} \tag{2.11}$$

For a planar vibration source, which radiates energy uniformly, from Equation (2.8) and Fig. 2.2b with $0.25E_o$ at each point:

$$\frac{\Delta_{wr}}{\Delta t} = \sqrt{\frac{3 \cdot E_o}{8 \cdot \rho \cdot \pi} \cdot \sum_{i=1}^{4} \frac{1}{r_i^3 \cdot e^{k \cdot r_i}}} \tag{2.12}$$

For a prismatic vibration source, which radiates energy uniformly, from Equation (2.8) and Fig. 2.2c with $0.125E_o$ at each point:

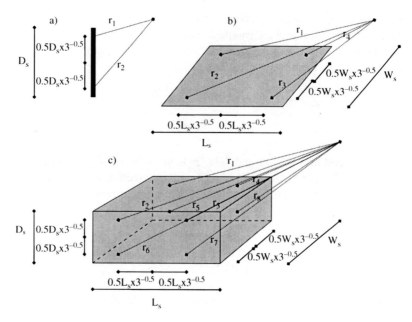

Fig. 2.2 (**a**) Linear, (**b**) planar, and (**c**) prismatic vibration sources with locations of Gauss integration points (e.g. Zienkiewich and Taylor, 1991)

$$\frac{\Delta_{wr}}{\Delta t} = \sqrt{\frac{3 \cdot E_o}{16 \cdot \rho \cdot \pi} \cdot \sum_{i=1}^{8} \frac{1}{r_i^3 \cdot e^{k \cdot r_i}}} \qquad (2.13)$$

2.3.2 Surface Waves

Surface (i.e. near surface) waves originate from shallow vibration sources such as described in Sections 1.3.1.2, 1.3.1.3, 1.3.2 and 1.3.3 or from body waves when they arrive at the surface. Two types of surface waves are of engineering significance

- Rayleigh waves are similar to water ripples except that ground particles move in opposite direction in the case of Rayleigh waves unlike water waves, which particles rotate as wheels on cars do. The Rayleigh wave velocity is similar to, but slightly lower than the transversal wave velocity. The minimum horizontal distance r_f at which Reyleigh waves appear at the surface from body waves is (e.g. Kramer, 1996)

$$r_f = \frac{D_s}{\sqrt{\left(\frac{v_l}{v_r}\right)^2 - 1}}, \qquad (2.14)$$

where D_s is the depth of a vibration source, v_l and v_r are velocities of the longitudinal and Rayleigh wave respectively.

- Love waves can develop if a ground layer with lower body wave velocity exist under the ground surface. Love waves propagate near ground surface in the horizontal direction similar to a snake movement due to multiple reflection of horizontal body transversal waves that are trapped within the subsurface layer. Love wave velocity range from the transversal wave velocity of the half space to the transversal wave velocity of the subsurface layer (e.g. Kramer, 1996).

The maximum amplitude Δ_w of ground displacement in the near field at a distance $r \leq r_f$ from a vibration source with the maximum force amplitude P_f at the ground surface is defined by the following Green function (e.g. Wolf, 1994)

$$\Delta_w = \frac{1-v}{2 \cdot \pi \cdot G \cdot r} \cdot P_f, \tag{2.15}$$

and in the far field at a distance $r > r_f$

$$\Delta_w = \frac{1-v}{2 \cdot \pi \cdot G \cdot \sqrt{r_f \cdot r}} \cdot P_f, \tag{2.16}$$

where v is Poisson's ratio, G is ground shear modulus. The peak ground velocity $\Delta_w \Delta t^{-1} = \Delta_w 2\pi f$, where f is the (predominant) vibration frequency. Wolf (1994) defined distance r_f from a point source on the ground surface to the near field as

$$r_f = \frac{2 \cdot \pi \cdot v_r}{\omega} \cdot \left(\frac{1-v}{2 \cdot \pi}\right)^2$$

$$\cdot \frac{\left\{\frac{8 \cdot v_t}{v_r} - \left[48 - 32 \cdot \left(\frac{v_t}{v_l}\right)^2\right] \cdot \left(\frac{v_t}{v_r}\right)^3 + 48 \cdot \left[1 - \left(\frac{v_t}{v_l}\right)^2\right] \cdot \left(\frac{v_t}{v_r}\right)^5\right\}^2}{\left[2 \cdot \left(\frac{v_t}{v_r}\right)^2 - 1\right]^4 \cdot \left[\left(\frac{v_t}{v_r}\right)^2 - \left(\frac{v_t}{v_l}\right)^2\right]}, \tag{2.17}$$

where ω is wave circular frequency, v is Poisson's ratio, v_t, v_l, v_r are velocities of the transversal, longitudinal and Reyleigh waves respectively.

For a linear vibration source at/near the ground surface, which radiates energy uniformly at the ground surface, the maximum velocity $\Delta_w (\Delta t)^{-1}$ in the near field at a distance $r \leq r_f$ from a vibration source with the maximum force amplitude P_f at/near the ground surface is using two Gauss integration points shown in Fig. 2.2a with $0.5 P_f$ each:

$$\frac{\Delta_w}{\Delta t} = \frac{f \cdot (1-v)}{G} \cdot \frac{P_f}{2} \cdot \sum_{i=1}^{2} \frac{1}{r_i}, \tag{2.18}$$

and in the far field at a distance $r > r_f$

$$\frac{\Delta_w}{\Delta t} = \frac{f.(1-v)}{G} \cdot \frac{P_f}{2} \cdot \sum_{i=1}^{2} \frac{1}{\sqrt{r_f \cdot r}} \tag{2.19}$$

For a planar vibration source at/near the ground surface, which radiates energy uniformly at the ground surface, the maximum velocity $\Delta_w(\Delta t)^{-1}$ in the near field at a distance $r \le r_f$ from a vibration source with the maximum force amplitude P_f at/near the ground surface is using four Gauss integration points shown in Fig. 2.2b with $0.25 P_f$ each:

$$\frac{\Delta_w}{\Delta t} = \frac{f.(1-v)}{G} \cdot \frac{P_f}{4} \cdot \sum_{i=1}^{4} \frac{1}{r_i}, \tag{2.20}$$

and in the far field at a distance $r > r_f$

$$\frac{\Delta_w}{\Delta t} = \frac{f.(1-v)}{G} \cdot \frac{P_f}{4} \cdot \sum_{i=1}^{4} \frac{1}{\sqrt{r_f \cdot r}} \tag{2.21}$$

For a prismatic vibration source near the ground surface, which radiates energy uniformly at the ground surface, the maximum velocity $\Delta_w(\Delta t)^{-1}$ in the near field at a distance $r \le r_f$ from a vibration source with the maximum force amplitude P_f near the ground surface is using eight Gauss integration points shown in Fig. 2.2c with $0.125 P_f$ each:

$$\frac{\Delta_w}{\Delta t} = \frac{f.(1-v)}{G} \cdot \frac{P_f}{8} \cdot \sum_{i=1}^{8} \frac{1}{r_i}, \tag{2.22}$$

and in the far field at a distance $r > r_f$

$$\frac{\Delta_w}{\Delta t} = \frac{f.(1-v)}{G} \cdot \frac{P_f}{8} \cdot \sum_{i=1}^{8} \frac{1}{\sqrt{r_f \cdot r}} \tag{2.23}$$

A vibration source is considered to be near the ground surface when it does not induce body waves, which propagate towards the ground surface.

2.4 Ground Wave Path Effects and Other Influential Factors

2.4.1 Impedance

Increase in the amplitudes of waves when they propagate into media of lower density ρ_1 and wave propagation velocity v_1 (usually towards surface) can be explained by considering the principle of conservation of energy and the energy-flux density per unit of time in the direction of wave propagation (upwards).

$$\frac{1}{2} \cdot v_1 \cdot \rho_1 \cdot \Delta_{w1}^2 \cdot \omega^2 \cdot dS = \frac{1}{2} \cdot v_2 \cdot \rho_2 \cdot \Delta_{w2}^2 \cdot \omega^2 \cdot dS \; from \; which$$

$$\frac{\Delta_{w1}}{\Delta_{w2}} = \sqrt{\frac{\rho_2 \cdot v_2}{\rho_1 \cdot v_1}}, \tag{2.24}$$

where ω is circular frequency, the product ρv is called soil impedance ($\rho_2 v_2$) ($\rho_1 v_1$)$^{-1}$ is called the impedance contrast between two adjacent layers. Equation (2.24) is applicable to the case of an isotropic stress-strain relationship in a non-dispersive (closed) system. In the case of greater ground strain, increased material damping causes the energy of propagating waves upwards to be partly transformed into heat. This energy loss causes a decrease in the difference between the amplitudes Δ_{w1} and Δ_{w2}.

2.4.2 Refraction

Using Fermat's principle, Snell showed (e.g. Kramer, 1996) that the ratio between sine of the angle between the wave path and the normal to the interface between two layers and the velocity of longitudinal or transversal waves is constant. Snell's law indicates that waves travelling from higher velocity materials into lower velocity materials will be refracted closer to the normal of the interface (Fig. 2.3) and vice versa.

Wave refraction affects also wave amplitudes because the source to location distance becomes shorter in layered ground than in a homogeneous ground as indicated in Fig. 2.4. The path length effect is more important in the near field than far field from a vibration source.

It should be mentioned that incident longitudinal waves refract (and reflect) as both longitudinal and vertical transversal waves and equally incident vertical transversal waves refract (and reflect) as both vertical transversal and longitudinal waves (e.g. Kramer, 1996).

Simplified analyses are based on homogeneous soil and therefore layered soil needs to be represented by an equivalent homogeneous soil. Average velocity of an equivalent homogeneous soil in the vertical direction $\bar{v}$ is

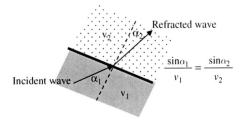

Fig. 2.3 Constant ratios between the sine of angle to transversal wave velocities according to Snell's law

Fig. 2.4 Effect of wave
refraction on the wave travel
path and wave amplitude

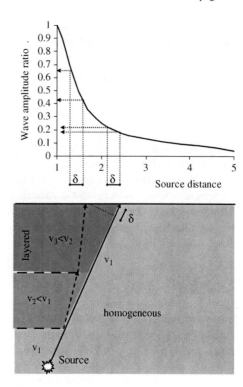

$$\bar{v} = \frac{\sum d_i}{\sum \frac{d_i}{v_i}}, \qquad (2.25)$$

When geometry is considered instead of velocity like in Equations (2.11), (2.12), (2.13), and (2.18), (2.19), (2.20), (2.21), (2.22), (2.23) then the source depth D_s in an equivalent homogeneous ground with wave velocity v_1 is calculated as

$$D_s = d_1 + \sum \frac{v_1}{v_i} \cdot d_i, \qquad (2.26)$$

where the sum is taken down to actual source depth, v_1, v_i are ground wave velocities in the first and ith ground layer with thickness d_i.

2.4.3 Reflection

According to Huygens-Fresnel principle (e.g. http://en.wikipedia.org/wiki/Huygens-Fresnel_principle) that each point subjected to vibration effect becomes a source of vibration itself it follows that ground waves not only refract but also reflect from boundaries of layers with different stiffness and ground wave velocities. The angle of reflected wave equals the angle of incident wave (Fig. 2.5) according to Fermat's principle of the least time for wave reflection.

Depending on the properties of the layers on two sides of boundary between them, the ratios between axial transmitted σ_T, reflected σ_r and incident σ_I stresses are (e.g. Dowding, 2000)

$$\frac{\sigma_T}{\sigma_I} = \frac{2 \cdot \rho_2 \cdot v_2}{\rho_2 \cdot v_2 + \rho_1 \cdot v_1}$$

$$\frac{\sigma_R}{\sigma_I} = \frac{\rho_2 \cdot v_2 - \rho_1 \cdot v_1}{\rho_2 \cdot v_2 + \rho_1 \cdot v_1}, \tag{2.27}$$

where subscripts 1 and 2 correspond to incoming and transmitting layers respectively with unit densities $\rho_{1,2}$ and wave velocities $v_{1,2}$. From Equation (2.27) it follows that a compressive wave reflects as compressive wave from a fixed boundary (bedrock) and as tensile wave from a free boundary. Non-cohesive ground cannot sustain tension and therefore reflected wave causes it's loosening. It should also be mentioned that transversal waves reflect from a fixed boundary (bedrock) with the same amplitude as the amplitude of the incoming wave but of opposite sign. At free boundary, the shear stress must be zero and incoming transversal wave doubles its incoming amplitude (e.g. Dowding, 2000), Fig. 2.6.

The ratios between the amplitudes of axial reflected Δ_{wR}, transmitted Δ_{wT} and incident Δ_{wI} waves are calculated from the following equations (e.g. Kramer, 1996)

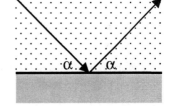

Fig. 2.5 Equal angles of incoming and reflected wave

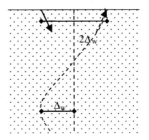

Fig. 2.6 Doubling of wave amplitude Δ_w at a free boundary

$$\frac{\Delta_{wR}}{\Delta_{wI}} = \frac{1 - \rho_2 \cdot v_2/\rho_1 \cdot v_1}{1 + \rho_2 \cdot v_2/\rho_1 \cdot v_1}$$
$$\frac{\Delta_{wT}}{\Delta_{wI}} = \frac{2}{1 + \rho_2 \cdot v_2/\rho_1 \cdot v_1},$$

(2.28)

where subscripts 1 and 2 correspond to incoming and transmitting layers respectively with unit densities $\rho_{1,2}$ and wave velocities $v_{1,2}$.

2.4.4 Superposition and Focusing

Superposition of incoming and reflected waves (from boundaries between ground layers with different stiffness or because of inertial structural interaction) could cause both wave amplitude increase and decrease depending if the waves are in phase or out of phase, Fig. 2.7. Obviously, wave amplitude increase is undesirable and need to be avoided or minimised whenever possible.

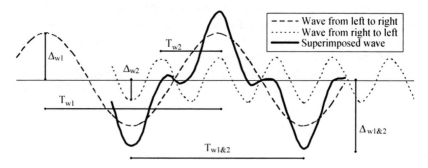

Fig. 2.7 An example of wave amplitude superposition

Curved boundaries of ground layers are capable of focusing (grouping) of ground waves where their amplitudes may be superimposed. Visual example is focused light by an optical lens. Curved boundaries of bedrock within sediment basins may focus and cause amplification of ground waves because of their superposition. The cases of wave focusing may not be frequent but the consequences may be serious and therefore need to be avoided or minimised whenever possible.

2.4.5 Ground Stiffness and Its Anisotropy

The correlations between ground axial stiffness i.e. Young modulus E and shear stiffness i.e. shear modulus G and ground longitudinal v_l and transversal v_t wave velocities are given in literature (e.g. Dowding, 2000) as

$$E = \rho \cdot v_l^2$$
$$G = \rho \cdot v_t^2$$
$$G = \frac{E}{2 \cdot (1 + v)},$$

(2.29)

where v is Poisson's ratio. When unit soil density ρ is in kg/m^3 and ground wave velocity is in m/s then modules E and G are in N/m^2 i.e. Pascal. Besides inherent ground anisotropy by formation, induced soil anisotropy is a result of different loading/unloading in the horizontal and vertical direction (e.g. Tatsuoka et al. 1997; Hashiguchi, 2001). Examples of processes causing inherited anisotropy of soil are sedimentation in lakes and seas and wind transportation and deposition while those causing induced anisotropy in soil are over consolidation i.e. removal by erosion of over laying soil deposits and melting of glaciers (from the last ace-age). Apparent soil over consolidation can be caused by desiccation, cementation and secondary consolidation. Examples of processes causing induced anisotropy of rocks are sedimentation and metamorphism while those causing induced anisotropy are folding and faulting.

Potts and Zdravkovic (1999), for example, indicated that the ratio between the vertical and horizontal Young modulus can reach up to 2 in clay, silt and sand. As Young and shear modulus are proportional it follows that the ratio between the vertical and horizontal shear modulus can reach up to 2 as well. Gerrard (1977), for example, tabulated a large number of data for soil and rocks and indicated that the ratio between vertical and horizontal Young modulus, as well as the shear modulus, could reach up to 5 in rocks and between 0.5 and 2 in soil.

The consequence of ground anisotropy is that the fronts of wave propagation will not be spheres but ellipsoids elongated in the vertical direction when $E_v E_h^{-1} > 1$. If an anisotropic case is considered as an equivalent isotropic case then the vertical distances need to be multiplied by the ratio between the horizontal and vertical wave velocities i.e. by the ratio $(E_h E_v^{-1})^{1/2}$.

2.4.6 Geometric (Radiation) Damping

Geometric (radiation) damping occurs due to spread of stresses and energy by waves from a source. The principle of conservation (preservation) of energy in an elastic system (e.g. http://en.wikipedia.org/wiki/Conservation_of_energy) states that energy remain constant in an isolated system and cannot be created or destroyed. As wave front spreads through ground or along a surface, the wave amplitudes have to decrease because the total energy along the front surface or circumference must remain the same. Similar applies to stress waves, as the surface on which the stresses act becomes bigger with distance the stress magnitudes must decrease, Fig. 2.8.

Geometric damping is an efficient mechanism for decreasing of amplitude of ground waves. Expressions for variation of wave amplitudes due to radiation damping (with distance) are given in Section 2.3.

Fig. 2.8 Wave/stress
amplitude (*arrows*) decrease
with wave/stress front
(*segment*) spread

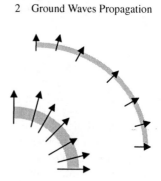

2.4.7 Material Damping

Material damping causes wave energy loss by its transformation into heating
because of friction between ground particles with wave propagation. Material damping is strongly dependent on amount of ground deformation. At small ground
deformation (shear strain of about 0.0001%), material damping is only a part of
a percent in rock and a few percent in soil. This amount of material damping can be
determined from measurement of wave amplitude decay with distance as shown in
Fig. 2.9.

   In one-dimensional wave propagation case, the radiation damping does not exist.
In this case, and from Equation (2.7) it follows that the attenuation coefficient κ is

$$\kappa = \frac{1}{r_s} \cdot \ln_e \frac{\Delta_{wo}}{\Delta_{wr}}, \tag{2.30}$$

where r_s is the distance where the displacement amplitudes Δ_{wo} and Δ_{wr} have been
measured in laboratory. Material damping is expressed also in terms of damping
ratio at large shear strain. Damping ratio is the ratio between actual and critical
damping, which prevents ground vibration.

   Laboratory tests have shown that soil stiffness and soil damping (energy dissipation) at large shear strain is influenced by cyclic strain amplitude, density and
acting mean principal effective stress of coarse grained soil, plasticity index and
over consolidation ratio of fine grained soil, and number of loading cycles (e.g. Seed
and Idriss, 1970; Hardin and Drnevich, 1972; Vucetic and Dobry, 1991; Ishibashi,
1992, etc.). Effective stress is called the difference between total stress and pore

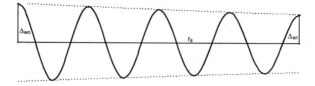

Fig. 2.9 Amplitudes Δ_{wo} and Δ_{wr} of a one-dimensional harmonic wave

Fig. 2.10 A hysteretic loop
in one cycle of soil shearing
at large shear strain

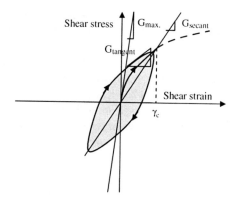

water pressure. A typical relationship between applied shear stress and induced
shear strain, within soil under symmetric cyclic loading, exhibits a hysteretic loop
as sketched in Fig. 2.10.

A shear modulus G is a ratio between incremental shear stress and shear strain.
Several different measures of shear modulus are shown in Fig. 2.10. When the increments are related to the origin (zero values) then so called secant modulus G_{secant} is
obtained. If the increments are related to the change in values from previous values
then the tangent modulus $G_{tangent}$ is obtained. Shear modulus dependence on shear
strain amplitude and other factors is determined by laboratory tests (e.g. ASTM
D4015; ASTM D3999) or from formulae (e.g., Zhang et al., 2005).

With an increase in shear strain, slippage between grains causes a weakening of
the soil structure, and a decrease of its shear strength and stiffness. This process
results in rotation of hysteretic loop towards horizontal axis. It should be noted that
the curve shown in Fig. 2.10 is for one cycle of loading/unloading. The curve for
greater number of cycles may change if soil strength and stiffness change (decrease)
with increase in number of cycles or with excess pore water pressure increase. For
idealised linear elastic materials, the hysteretic loop and the backbone curve are
straight and coincidental lines.

At very small shear strains, less than about 0.0001%, hysteresis is virtually
absent, and the behaviour of the soil is often approximated as linear-elastic. When
the soil indeed behaves as an isotropic linear elastic body, the shear modulus
G_{max} is

$$G_{max.} = \rho \cdot v_t^2, \tag{2.31}$$

where ρ is soil unit density (kg/m^3) and v_t is soil transversal wave velocity. The latter can be determined using field geophysical methods (e.g. ASTM D4428; ASTM
D5777) or seismic cone (e.g. Lunne et al., 2001) to avoid possible problems caused
by sample disturbance, size and orientation effects and measurements of very small
strain in the laboratory. Disturbance of loose sample causes their artificial compaction and of dense samples their artificial loosening. Sample disturbance can be

minimized by using thin walled soil samplers, pushing instead of hammering of soil samplers and by rapid freezing of soil before sampling. The later technique is very expensive and not widely used in practice. Several researches (e.g. Hardin, 1978; Seed and Idriss, 1970) suggested formulae for calculation of G_{max} based on empirical correlations with other soil properties.

The width of the hysteretic loop is related to the area, which is a measure of internal energy dissipation. The dissipation involves the transformation of energy or work into heat, by particles friction due to their movements. A damping ratio ξ is frequently used as a measure of the energy dissipation (e.g. Kramer, 1996).

$$\xi = \frac{E_D}{4 \cdot \pi \cdot E_{max}} = \frac{1}{2 \cdot \pi} \cdot \frac{A_{loop}}{G_{secant} \cdot \gamma_c^2}, \qquad (2.32)$$

where E_D is the dissipated energy, E_{max} is the maximum strain energy, i.e. the area of the triangle in Fig. 2.10 bordered by G_{secant} line, the vertical at γ_c and shear strain axis; and A_{loop} is the area of the hysteretic loop. Soil parameters G_{secant} and ξ are often referred to as equivalent linear soil parameters. Soil damping at large strain is determined by laboratory tests (e.g. ASTM D3999, D4015) or from formulae (e.g., Zhang et al., 2005). The importance of damping ratio on response spectrum is shown in Section 4.4.3. The frequency of damped vibration is only $(1-\xi^2)^{1/2}$ times smaller than the frequency of an undamped vibration (e.g. Kramer, 1996). For usually small values of ξ the difference is not significant.

Shapes of shear modulus and damping ratio functions versus logarithm of shear strain for Quaternary formations and effective confining stress equal to the atmospheric pressure are shown in Fig. 2.11, for example.

Soil damping is sometimes represented by a viscous coefficient for calculation convenience. Soil viscosity coefficient η is related to the damping ratio ξ (Equation 2.32) as $\eta = G \cdot \xi / (\pi \cdot f)$, where G is shear modulus; f is frequency of vibration (e.g. Kramer, 1996). From the expression for damping ratio ξ (e.g. Thompson, 1965)

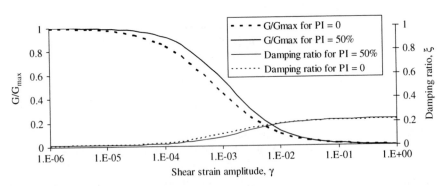

Fig. 2.11 Typical modulus G/G_{max} and damping ξ ratio versus logarithm of shear strain γ for quaternary soil

$$\xi = \frac{1}{2 \cdot \pi} \cdot \ln_e \frac{\Delta_{wo}}{\Delta_{wr}}, \tag{2.33}$$

and Equation (2.30) it follows that $\xi = \kappa r_s (2\pi)^{-1}$.

For a uniform isotropic soil layer subjected to a harmonic horizontal motion from a vibration source of large planar or prismatic shape, the amplification factor of the amplitudes of at depth acceleration $a_{peak,depth}$ and at the surface $a_{peak,surface}$ above a large vibration source is for Kelvin-Voigt soil model (e.g. Kramer, 1996);

$$\frac{a_{peak,surface}}{a_{peak,depth}} = \frac{1}{\sqrt{\cos^2\left(\omega \cdot H/v_t\right) + \left[\xi \cdot \left(\omega \cdot H/v_t\right)\right]^2}}, \tag{2.34}$$

where ω is the circular frequency of ground shaking, H is the layer thickness, v_t is transversal wave velocity and ξ is damping ratio. The base acceleration amplification factor is shown in Fig. 2.12.

From Fig. 2.12 it follows that the amplification factor and its scatter are largest at smaller frequencies of ground shaking and soil damping ratio but also that they could reach large values even for greater damping ratios in linear elastic soil.

2.4.7.1   Example of the Effect of Material Damping on Peak Particle Velocities

An example of the effect of material damping ratio on peak particle velocity is provided using Equation (2.8) and $E_o = 1$, $\rho = 1900$ kg/m^3. Calculated peak particle velocities versus distances from a vibration source of unit energy are shown in Fig. 2.13.

From Fig. 2.13 it follows that damping ratio has greater effect at larger distances than at smaller distances (less than 10 m) between a vibration source and a receiver.

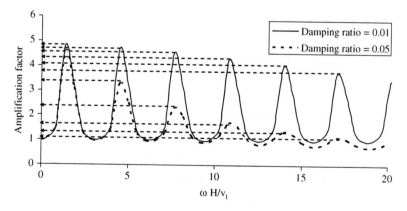

Fig. 2.12 Influence of frequency on amplification factor of a damped linear elastic layer

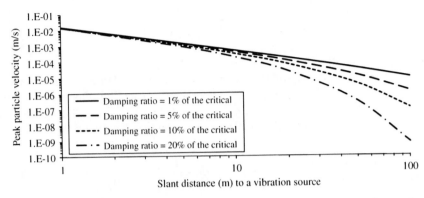

Fig. 2.13 Illustration of the effect of damping g ratio on peak particle velocity with distance from the source in Section 2.4.7.1

2.4.8 Soil Layering and Topography

Soils layering i.e. differences in ground wave velocities affect wave reflection and refraction phenomena. When an equivalent homogeneous soil is used in simplified analysis in place of layered soil then wave reflection and refraction can not be considered. The effect of using an equivalent homogenous soil in place of actual layers is considered in the following example for vertically propagating transversal waves.

2.4.8.1 Example of the Ratio Between Foundation and Ground Amplitudes for a Layered and an Equivalent Homogeneous Soil

Considered properties of a layered soil are given in Table 2.1.

An equivalent transversal wave velocity according to Equation (2.25) is $\bar{v} = \frac{30}{5/100 + 10/200 + 15/275} = 194$ m/s, an equivalent unit density is $(5 \times 1600 + 10 \times 1800 + 15 \times 1900) \times 30^{-1} = 1816$ kg/m^3 and an equivalent shear modulus is $1816 \times 194^2 \times 10^{-6} = 68.3$ GPa. Adopted equivalent Poisson's ratio is 0.3 and the damping coefficient 0.03.

The free-field ground motion amplitude for vertically propagating transversal waves with circular frequency ω is described by Wolf and Deeks (2004) as

Table 2.1 Soil properties considered in Section 2.4.8.1

Layer thickness (m)	Unit density (kg/m^3)	Transversal wave velocity (m/s)	Shear modulus G (GPa)	Poisson's ratio	Damping ratio ξ
5	1600	100	16	0.4	0.04
10	1800	200	72	0.3	0.03
15	1900	275	143.7	0.2	0.02

$$u^f(z, \omega) = u^f(\omega) \cdot \cos \frac{\omega}{v_t} \cdot z, \qquad (2.35)$$

where the depth z is measured downwards from the free surface, $u^f(\omega)$ is the surface amplitude of the free field ground motion, ω is the circular frequency of ground vibration, v_t is ground transversal wave velocity. Computer program CONAN (by Wolf and Deeks, 2004, http://w3.civil.uwa.edu.au/~deeks/conan/) is used for the calculations. Besides consideration of layered soil on rock base or a half space in the free-field, the horizontal, vertical and rocking vibration of foundations and circular cavities can be considered using an equivalent surface disk for a shallow foundation and stack of embedded equivalent disks for deep foundations and circular cavities.

The ratios between foundation and ground wave amplitudes at different frequencies of ground vibration considered in the example are shown in Fig. 2.14.

From Fig. 2.14 it follows that the differences in the amplitude vibrations of a foundation on the layered and its equivalent homogeneous soil are minimal in the case of vertically propagating shear waves. However, surface waves approaching an outcropping bedrock can experience increase in wave amplitudes similar to sea waves approaching a shore. Also the amplitudes of incoming and reflected surface

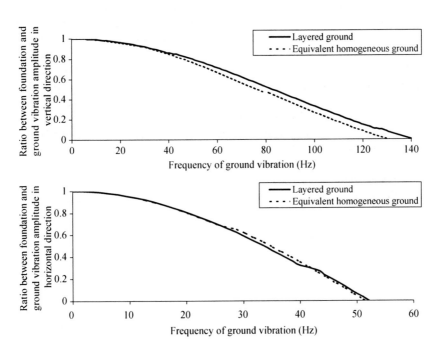

Fig. 2.14 Ratios in between foundation and ground vibration amplitudes in the vertical and horizontal direction in the Section 2.4.8.1

waves from an outcropping rock may be superimposed. The cases of extensive damages to the structures caused by seismic waves near the edges of sediment basins are well known and documented (e.g. Srbulov, 2008).

The cases of seismic wave amplification by slopes, ridges and canyons are known (e.g. Srbulov, 2008). There is no reason why ground waves caused by other sources than earthquakes are not also amplified by topography. The cases of artificial topographic effects i.e. trenches used for vibration isolation are considered in Section 8.3.

2.5 Summary

This chapter contains descriptions of types of ground waves and how amplitudes of ground waves can be determined, of ground wave path effects and of factors affecting propagation of ground waves.

- Body waves formed around deep wave sources could have large amplitudes if they are the results of blasting or due to installation of large displacement piles. Surface waves, which are formed when body waves arrive at the ground surface or by near surface vibration sources, usually have the greatest impact on the vibration receivers, which are described in Section 1.3.
- A number of ground wave path effects (impedance, refraction, reflection, superposition and focusing) can not be considered when an equivalent homogeneous isotropic half space is considered in simplified analysis instead of real ground. However, for nearly vertically propagating body waves, the effects of impedance, refraction, and reflection on the wave amplitudes is not significant as shown for one example in Section 2.4.8.1. Surface waves penetrate to shallow depth and for them the effects applicable to layered ground are less significant except in special circumstances as mentioned below.
- Of factors affecting propagation of ground waves, geometric (radiation) damping is more significant than material damping except in the near field of strong vibration such as caused by pile driving and blasting. Outcropping bedrock and substantial topography change could cause significant surface wave amplitude amplification but these cases are exceptional rather than usual occurrences.

Chapter 3
Ground Vibration Measurement

3.1 Introduction

Measurement of ground vibration is important for checking of amplitudes of predicted ground motion and for confirmation of efficiency of control measures of ground vibration. The properties of measuring instruments used can affect the results of measurements and therefore the user need to know how to select these properties in order to obtain correct results for a particular case of interest.

An essential part of a measuring instrument is its transducer (sensor), which converts the amplitude of particle motion into an electrical signal or light beam. The main characteristics of transducers in connection with their performance are (e.g. Dowding, 2000):

- Sensitivity i.e. the ratio between transducer's electrical outputs to its mechanical excitations (displacement/velocity/acceleration) for energy-converting transducers, which do not require an energy source for their operation. For passive transducers, which require an energy source for their operation, such as strain gauge or a piezo-resistive accelerometer, the sensitivity is specified in terms of output voltage per unit of measurement per unit of input voltage.
- Cross-axis or transverse sensitivity is the sensitivity to motion in a direction perpendicular to the direction of measurement. Some manufacturers refer to axis alignment, which reflects the extent to which the direction of measurement is parallel to the direction of instrument containing the transducer.
- Resolution is the smallest change in mechanical input that produces a recordable change in the electric output. Geophones have greater resolution than accelerometers at low frequencies.
- Frequency range over which the electrical output is constant with a constant mechanical input.
- Phase shift is the time delay between the mechanical input and the electrical output of the instrument.
- Calibration requirements are the allowable variations in electrical output with constant amplitude input when there are changes in frequency range of the input, change in environment, amplitude of the input, or time.

M. Srbulov, *Ground Vibration Engineering*, Geotechnical, Geological, and Earthquake Engineering 12, DOI 10.1007/978-90-481-9082-9_3, © Springer Science+Business Media B.V. 2010

- Environmental sensitivity describes the response of a transducer to humidity, temperature, or acoustic changes.
- Mass and size of a transducer are important when available space is limited.

Some transducer properties are exclusive. For example, a transducer with wider range of measurement may need to have smaller sensitivity and resolution in order to be able to operate at extreme excitation, or wider frequency range of measurement (bandwidth) may mean higher environmental sensitivity and more frequent re-calibration, etc. In other words, transducers have their limitations and, therefore, their applicability range.

Lucca (2003), for example, mentions the following problems arising from the limited applicability of transducer:

- Decoupling may happen when the coil inside of the magnetic field of a geophone moves large enough to disrupt the magnetic field and so exceeds the operational limits of the transducer when used for close-in monitoring near sources, such as pile driving and blasting. It also happens if instruments is not firmly attached to the base and moves more than it.
- Aliasing occurs whenever a signal from a transducer is not sampled at greater than twice the maximum frequency of the signal, which in effects means a sort of filtering of high frequencies and truncation of the maximum amplitudes of the vibration. To measure the amplitude correctly at high frequencies, the sample rate must be at least four to five times greater than the frequency of monitored vibration. Frequencies in the extreme near field can be as high as 6000 Hz. Although this signal attenuates quickly, it can still involve many hundreds of hertz within 6 m of a blast. This will cause exceeding of the operational limits of many commercial geophones and will generate erroneous data.

Bormann (2002) highlighted one important property of vibration data acquisition and processing

- High signal to noise ratio to avoid sometimes complete masking of ground vibration caused by a particular vibration source by environmental noise.

There are various modern technologies used to manufacture miniaturized sensors (bulk micro machining, surface micromachining, piezoelectric, piezo-resistive, thermal-mechanical, plated structures, etc.) that differ in terms of cost effectiveness and reliability in the applications requiring both high performance and robustness in operation.

The measurement systems also need an accurate method of storing data, adequate storage space, and transmitting data to a location where they can be analysed. Some systems download data by mobile phone at regular intervals, others have to be visited on site.

The objective of this chapter is to describe and comment on two commonly used types of instruments for measurement of ground velocity and acceleration.

3.2 Geophones

Geophone consists of a permanent magnet, coil, top and bottom springs, steel casing and cable connector. Geophone casing with cable connector is shown in Fig. 3.1.

The output from a velocity transducer (geophone) is generated by a coil moving through a permanent magnetic field. The voltage induced in the coil is directly proportional to the relative velocity between the coil and the magnetic field. The voltage output is usually high enough so that amplification is not required even if using long connecting sables. Velocity transducer response becomes nonlinear at low frequencies because it is a single degree of freedom oscillator (SDOFO), with the ratio between input and output velocities described by the following equation (e.g. Dowding, 2000)

$$\frac{v_c}{\partial \Delta w / \partial t} = \frac{\beta_t^2}{\sqrt{(1 - \beta_t^2)^2 + (2 \cdot \xi \cdot \beta_t)^2}},\tag{3.1}$$

where v_c is the relative velocity of the coil, $\partial \Delta_w (\partial t)^{-1}$ is ground velocity, $\beta_t = f_d f_o^{-1}$ is the tuning ratio, f_d is the frequency of an input motion i.e. ground, f_o is the frequency of the output motion i.e. transducer, ξ is the damping ratio (Section 2.4.7). The phase angle θ between v_c and $\partial \Delta_w (\partial t)^{-1}$ is (e.g. Dowding, 2000)

$$\theta = \arctan \left(\frac{2 \cdot \xi \cdot \beta_t}{1 - \beta_t^2} \right)\tag{3.2}$$

Figure 3.2 shows that a linear response to the excitation is best achieved for a damping ratio of about 0.6 and β_t greater than 2. For this reason, the natural frequency of a velocity sensor must be smaller than the smallest recorded frequency. The transducer frequency range is usually broadened by many manufacturers. Figure 3.2 shows that the change in phase angle θ decreases with the increase in β_t.

Basic geophone data are provided by the manufacturer together with the calibration certificates. Geophones may need to be re-calibrated either by the manufacturer or the user. Hiller and Crabb (2000) described how the calibrations of the geophones

Fig. 3.1 An example of uniaxial geophone casing about 30 mm high and 25 mm wide with cable connector at the top

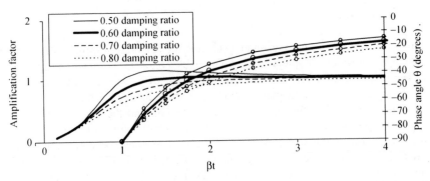

Fig. 3.2 Change of amplification factor and phase angle θ of a SDOFO oscillator with change in the tuning ratio β_t

were checked according to BS 6955 (1994): *A reference accelerometer was mounted back-to-back with each geophone in turn on an assembly mounted on an electro-dynamic vibration generator. The signals from the transducers were captured by the acquisition system and the peak particle velocity measured by each transducer was compared at frequencies of 15Hz and 70Hz. One vertical and two horizontal geophones were also tested over a range of frequencies from 4 to 300Hz to establish the frequency response curve for comparison with that supplied by the manufacturer. The sensitivity of each geophone was determined to be within the manufacturer's specific tolerance. As a part of the calibration procedure, the effect on the sensitivity of the geophones to misalignment was investigated, to assess whether it was necessary to measure the mounting alignment. This revealed that the deviation in sensitivity increased to 1% at ±4° tilt for a horizontal geophone and at ±13° for a vertical axis geophone. It was therefore concluded that alignment by eye would cause acceptably small error.* ISO 16063-16, 21 and 22 (2003) are used for the calibration of vibration and shock transducers internationally.

To ensure data integrity, all components of data acquisition system need to be calibrated before use. Signal amplifiers, filters, analogue to digital converters and recorders are calibrated by the manufacturers. The user need to check operational limits of equipment and expire date of the calibration before use.

Method of fixation of transducers to their bases is an important issue in ground vibration measurement. Individual transducers are usually screwed into three orthogonal faces of a metal cube to create triaxial array with one vertical and two horizontal components (e.g. Hiller and Crabb, 2000; ISO 4866, 1990). The arrays are then screwed on to say 200 mm long stainless steel spikes driven fully into ground. Where necessary, any loose soil or vegetation are removed before the spike are driven. If geophones are to remain in place for a long time and in order to minimize coupling distortion, they are buried to a depth at least three times the main dimension of the mounting unit. Each excavation is backfilled with the exca-vated soil hand-tamped around the array to ensure good coupling with the ground and to minimise the risk of disturbance. ISO 4866 (1990) states that, alternatively, transducers can be fixed to a rigid surface plate (for example a well-bedded paving slab).

The number of locations at which vibration measurements could be made simultaneously is restricted by the data acquisition system. Typically, geophones are positioned as close as practicable to a vibration source and then at distances oriented radially from the source depending on the source energy output. It is important to avoid decoupling and aliasing effects (Section 3.1), which are possible near blasting location and near driven piles. Lucca (2003) states that *If a seismograph is set up in an area where there are multitude of surfaces and structures, the interaction of the vibration waves with each other, surfaces and structures may cause the seismograph readings to be erroneous and not representative of the actual peak particle velocities affecting the structure.*

3.2.1 Short Period Sensors

Bormann (2002), for example, provides details of state of the art and practice in seismological observatory including vibration sensors. The short period sensors measure signals from approximately 0.1 to 100 Hz, with a corner frequency at 1 Hz of the sensor's frequency response function. They have a flat response to ground velocity for frequencies greater than this corner frequency. They are relatively stable in a broad range of temperatures. The electronic drift and mass position instability (usually associated with active sensors) are typically not a problem. There are active short period sensors, which are either electronically extended 4.5 Hz geophones or accelerometers with electronically generated velocity output. These sensors are often cheaper and smaller. Their drawback is that they require power and are more complicated to repair. Details of different sensors available on the market are given in volume 2, DS 5.1 by Bormann (2002).

Short period vibration noise has an important effect on the measurements. Natural sources of short period noise are wind (i.e. vibrations in the range from 0.5 Hz up to 60 Hz caused by air friction over rough terrain, trees or objects), rushing water (waterfalls or vortexes in streams). Dominant sources of high frequency noise are man made (rotating or hammering machinery, road and rail traffic, etc.). Most of these sources are distributed, stationary or moving, coming from various directions to superpose to a rather complex, more or less stationary random noise. Besides ambient noise, sensor's tilt, short term changes of ambient temperature and variation of atmospheric pressure also affect the measurements. Enclosure of sensor in an airtight housing greatly reduces the effects of variation of atmospheric pressure and also adiabatic changes of temperature. Short term changes of temperature are suppressed by the combination of thermal insulation and thermal inertia.

An example of the use of short-period sensors for micro-tremor field investigations into site effects in the town of Duzce, Turkey, by Tromans (2004) follows.

3.2.1.1 Case Study of Micro-tremor Field Investigation into Site Effects in Duzce – Turkey by Tromans (2004)

Fourier amplitude spectrum (FAS) (Section 4.4.1) has limited use in the estimation of site response. The spectral amplitudes are often dominated by source and path effects, which are not easily decoupled from the ground motion. Bard (1998) reviewed recent studies and suggests that absolute FAS can reflect site response if one or both of the following are fulfilled:

- Impedance contrast (Section 2.4.1) between soil and bedrock at depth is high. This leads to trapping of surface and/or body waves giving rise to a conspicuous spectral peak at the resonant frequency.
- Soil layers are very deep causing a low fundamental frequency.

A number of researchers were able to identify the soil fundamental frequency from FAS of micro-tremor records due to presence of very soft soil. To solve the problem in other cases, the ratio of horizontal to vertical Fourier amplitude spectra (HVSR) method was first introduced by Nogoshi and Igarashi (1971). It was subsequently used by Nakamura (1989) and has since been extensively used for the mapping of site effects due to its ability to estimate site effects without the need for simultaneous reference site measurements. The HVSR has been effective for the estimation of fundamental frequency and even spectral amplification factors in different grounds. In spite of its popularity, a satisfactory theoretical explanation has not been agreed (Tromans, 2004).

Micro-tremor investigation of site effects requires that measurements are taken at many locations for a few minutes. Besides sensitivity, portability and ease of installation are therefore the most important instrument characteristics. The set up used to measure ambient noise consists of three orthogonal miniature sensors with 1 Hz frequency and a flat frequency response over the range 1–80 Hz, Fig. 3.3. A series of measurements performed using a more sensitive long period (30 s) seismometer was carried out to verify the reliable range of the short period instrument. The instrument mass is 7.5 kg and power consumption is 3 W. The instrument only

Fig. 3.3 Set up to measure ambient noise at a rock site just outside Duzce (Tromans, 2004)

requires levelling with up to ±2.5° tilt but not sensor mass unlocking or mass centring prior to each reading. Data acquisition can begin as soon as the sensor has been powered up.

Two sets of equipment were used to speed up data collection, reduce the risk of instrument downtime in the field and to allow simultaneous monitoring at selected sites to investigate the effects of various factors on the stability of HVSR calculated from the micro-tremor data. True reference site measurements were considered unfeasible for Duzce due to the lack of any suitable rock reference site within a distance appropriate for the frequencies of interest. Bard (1998) stated that soil site to reference site distance for frequencies greater than 1 Hz, which is within the range of interest for the Duzce study, should not exceed 500 m, unlike the distance between the centre of Duzce and the nearest rock site of about 6 km. Batteries were selected to ensure adequate power supply for the estimated maximum number of daily measurements, whilst minimising weight. Overnight battery recharging from the mains and emergency in-car charging was available. Data backup was carried out every night using a portable CD writer.

A set of site selection guidelines are listed as follows:

- **Maintain consistent instrument-ground interface**. Asphalt or concrete surfaces are avoided due to their unknown effects on the ground response. Compact soil is considered the best interface, offering a direct link to the soil and a level surface for easy set up. Where found, loose stones are swept away to enable a stable contact between instrument and soil. Areas of ground with significant cracking are avoided. A crack creates a vertical free surface, which could affect characteristic of ground vibration. Locations immediately adjacent to ditches or cuttings are also to be avoided for the same reason.
- **Avoid unusual subsurface conditions**. These include buried basements, sewer or water pipes etc. Longer established areas of open space (e.g. road verges, gardens, fields, parks) ere selected in preference to areas of recently disturbed ground.
- **Avoid excessively windy conditions**. Wind acting on the seismometer can cause distortions in the horizontal components of motion at frequencies below about 1.5 Hz.
- **Avoid rainy condition or wet ground**. Rain falling on or near the instrument would disturb measurements. Also, increased water content of surface soil layers could modify ground response.
- **Avoid strong local sources of noise**. No measurements were taken immediately adjacent to main roads to avoid strong, persistent transient vibrations caused by heavy traffic.

For consistency, the same setting up and measurement procedure was followed for each micro-tremor location as follows:

- **Instrument set up**. Level instrument, trail cables, with slack to minimise spurious vibrations. Set up laptop computer as far as possible from instrument to minimise user induced ground disturbance. Initiate data acquisition software and then power up instrument.

- **Field notes**. Log instrument ID and location coordinates (from GPS), draw a simple schematic location plan highlighting significant local features and to enable repeatability of measurements. Record any significant event or unusual disturbance (e.g. pedestrians, livestock, vehicles, machinery).
- **Data monitoring**. Monitor data in real time using the manufacturer software to identify any instrument malfunction or unusual disturbance.

Following some basic pre-processing, Fourier spectra and HVSR were computed for measurements made at each location using MATLAB software as follows:

- Read samples and header information.
- Convert data from bits to velocity.
- Correct zero offset.
- Apply window function to samples vector.
- Perform Fast Fourier Transform.
- Calculate smoothed FAS, instrument corrected.
- Loop for each component.
- Combine the horizontal spectra prior to calculation of the HVSR.
- Calculate average horizontal to vertical spectral ratio for single time window.
- Loop for each time window.
- Calculate average HVSR for all time windows.
- Plot spectra.
- Determine predominant frequency where HVSR has the greatest amplitude.

An example plot of horizontal to vertical spectral ratios (HVSR) is shown in Fig. 3.4.

A total of 121 micro-tremor measurements were made in and around Duzce over a period of about two weeks during May and June 2001. Most measurements were made in the central part of the town. For each location, specified by the point ID, GPS coordinates, instrument ID and measurement time and date are recorded. Field notes made during measurements were summarized into three columns: ground contact, location category and location description.

Comparisons between the results of measurements of two instruments set up adjacent to each other at a quit semi-rural site conformed the consistency of micro-tremor measurements over the whole data set. Simultaneous ambient noise measurements were made with the 1 and 30 s instruments at Duzce strong motion station, DZC. Spectral ratios between the two instruments were calculated from FAS for each of 5 sample windows of 40.96 s duration. The spectral ratios are obtained to follow the instrument transfer function down to about 0.2 Hz. This confirms that Fourier amplitudes obtained from the 1 s instruments are reliable down to 0.2 Hz, which in any case is the frequency limit defined by the smoothing algorithm employed. Reliable result may be possible at even lower frequencies by using a narrower smoothing bandwidth.

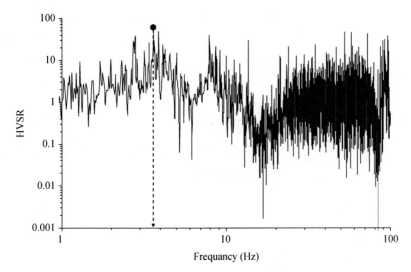

Fig. 3.4 An example plot of horizontal to vertical spectral ratios (HVSR) from the aftershock of the Izmit earthquake with recorded 0.053 g peak horizontal acceleration at Duzce strong motion station on 7 July 2000 (from data by Ambraseys et al., 2004)

Amplitudes of FAS are obtained to vary significantly from site to site in and around Duzce. For spectra from any pair of sites, the variation will be due to differences in both noise source characteristics and site conditions. At each location, spectra can also vary significantly with time due to the change in ambient noise levels during day. The increased noise levels affected spectral amplitudes from 0.5 Hz upwards, with noticeable differences between 3 and 10 Hz. In this frequency band, the amplitudes of the noisier samples are up to an order of magnitude greater. The variations in HVSR between the two sets of measurements are much less than variations in FAS as observed in many other micro-tremor studies (e.g. Bard, 1998). In the exceptional case in Duzce, the increased noise levels appear to have increased the amplitude of the predominant peak by about 40% and reduced its frequency by about 25%.

For most of the measurements taken in Duzce, wind was negligible. However, several data points had to be excluded from the subsequent analyses due to adverse effects of wind on HVSR amplitudes, indicated by obscured HVSR peak below about 1.5 Hz, in the frequency range of interest. HVSR amplitudes are augmented because wind affects the horizontal components of FAS more than the vertical components.

The HVSR of earthquake records from DZC strong motion station confirm the predominant frequency obtained from micro-tremor measurements made at the same location. No consistent correlation has been found between the amplitudes obtained using the two approaches. Results obtained at BOL strong motion station are less conclusive than at DZC station even for estimates of the predominant frequency.

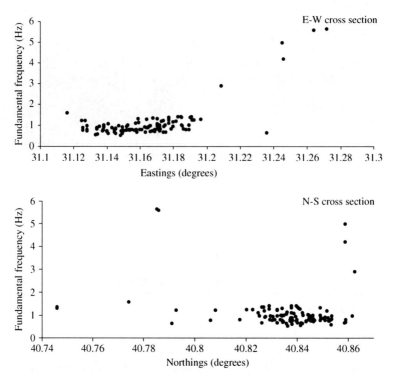

Fig. 3.5 Fundamental frequency of ground at Duzce – Turkey using data by Tromans (2004)

The fundamental ground frequencies inferred from the measurements are shown in Fig. 3.5.

Simsek and Dalgic (1997) provided a geological section in N-NE direction across the sediment basin under Duzce. The basin is about 18 km long and up to 250 m deep in the middle. Near the edges of the basin, the thickness of the sediment decreases to about 50 m. Because of such geological conditions, the outliers in Fig. 3.5 most likely represent sediments with different thicknesses and stiffness according to Equation (2.4) unless they are simply erroneous results. While the fundamental frequency of soil layers in Duzce town is about 3.6 Hz (Fig. 3.4) during an earthquake recorded at Duzce strong motion station located over about 210 m thick sediments, it is only about 1 Hz (Fig. 3.5) based on the measurements of ambient noise.

3.2.2 Long Period Sensors

Bormann (2002), for example, states that the broadband sensors are rather popular. They provide measurement of vibration from about 0.01 to 50 Hz and therefore allow a much broader range of studies than the short period sensors. However, the broad band sensors are more expensive and demand more efforts for installation and operation than short period sensors. Also, the broad band sensors require a higher level of expertise with respect to instrumentation and methods of analyses. They are active feedback sensors and require a stable power supply. They also require careful site selection, a better controlled environment and therefore are less robust than short period sensors. Because they do not attenuate the 0.12 to 0.3 Hz natural background noise peak, their raw output signal contains much more background noise than signals from a short period sensor. Useful vibration signals are often hidden by environmental noise and can be resolved and analyzed only after filtering to remove the background noise.

Very broad band sensors are utilized in global seismological studies. They are able to record the frequencies resulting from Earth's tides and free oscillations of the Earth. Their primary purpose is the research of the deep interior of the Earth. Their only important advantage over broad band sensors is their ability to record vibration at frequencies around and below 0.001 Hz. They are expensive, require very elaborate and expensive seismic shelters and rather difficult to install.

3.3 Accelerometers

Although measurement of particle velocity $\partial \Delta (\partial t)^{-1}$ by geophones directly yield ground stress as a product of ground unit density ρ, $\partial \Delta (\partial t)^{-1}$ and wave propagation velocity v, and ground strain as the ratio between particle velocity $\partial \Delta (\partial t)^{-1}$ and wave propagation velocity v (e.g. Dowding, 2000), geophone properties may not be suitable for the measurements in near field of high energy vibration sources.

Accelerometers are used for vibrations with amplitudes or frequencies outside the operating limits of geophones. Accelerometers have a range of ± 50 of the gravitational acceleration and a near linear response proportional to acceleration from about 1 Hz to 10 kHz, but are not suitable for low-frequency measurements when the outputs are integrated to obtain velocity. Further disadvantages of the accelerometers are that they require a power supply and are more susceptible to backgrounds noise than geophones (e.g. Hiller and Crabb, 2000). Bormann (2002) states *However, the latest generation accelerometers are nearly as sensitive as standard short-period (SP) seismometers and also have a large dynamic range. Consequently, for most traditional short period networks, accelerometers would work just as well as 1-Hz SP seismometers although the latter are cheaper. In terms of signal processing, there is no difference in using a seismometer or an accelerometer.*

A number of different types of accelerometers are available.

- Servo (or force balance) accelerometers use a suspended mass to which a displacement transducer is attached. When accelerometer housing is accelerated, the signal produced by the relative displacement between the housing and accelerometer mass is used to generate a restoring force that pushes the mass back towards its equilibrium position. The restoring force is proportional to the acceleration and can be measured electronically. Servo accelerometers can provide very good accuracy over the range of frequencies of greatest interest in earthquake engineering (e.g. Kramer, 1996). An example of a three componental force balance accelerometer casing is shown in Fig. 3.6.

- Piezoresistive accelerometers use piezoresistive strain gauges. The strain gauges are a solid state silicon resistors, which electrical resistance changes in proportion to applied stress and are small, a few centimetres. The upper limit of the frequency range is a few thousands hertz. They require an external power supply (e.g. Dowding, 2000). An example of a piezoresistive accelerometer casing is shown in Fig. 3.7.
- Piezoelectric accelerometers use the property of certain crystals to produce a voltage difference between their faces when deformed or subjected to a force. The accelerometer can work in compression or shear. The compression transducers are sensitive to the environment. The shear transducers are less affected by temperature changes, are lighter and have wider frequency response (e.g. Dowding, 2000). Examples of uniaxial piezoelectric accelerometer casings are shown in Fig. 3.8.
- Micro machined differential electric capacitive sensors are claimed to offer a lower noise floor and significantly better stability performance over time and over temperature then piezoresistive accelerometers. The change of the sensor element capacity with acceleration is detected and transformed into voltage by a converter, which includes an amplifier and an additional integrator that forms an electronic feedback circuit. An example of a uniaxial micro machine capacitive acceleration sensor is shown in Fig. 3.9.

Besides sensors and accelerometer casing, an important component of an accelerometer is an accurate clock, particularly when more than one component of motion is measured or when the ground motion at different locations are compared. Modern instruments maintain time accuracy by synchronizing on daily bases with

Fig. 3.6 An example of a three componental force balance accelerometer casing about 6 cm high and 13 cm diameter

Fig. 3.7 An example of a uniaxial piezoresistive accelerometer casing about 15 mm in diameter and 25 mm high

Fig. 3.8 An example of uniaxial piezoelectric accelerometer casings a few millimetres large

Fig. 3.9 An example of a micro machine capacitive acceleration sensor about 10 mm wide

radio time signals transmitted by a standard time service or recording such signals along with the ground motion data. Universal Coordinated Time (i.e. Greenwich Mean Time) is used as a common worldwide basis (Kramer, 1996). Some micro machined sensors are also supplied with Global Positioning System receiver.

For a single degree of freedom oscillator (SDOFO) representing an accelerometer, the ratio between absolute values of the input acceleration and the output displacement is described by following equation (e.g. Dowding, 2000)

$$\frac{|\delta_t|}{|\partial^2 \Delta_w / \partial t^2|} = \frac{1}{(2 \cdot \pi \cdot f_o)^2} \frac{1}{\sqrt{(1 - \beta_t^2)^2 + (2 \cdot \xi \cdot \beta_t)^2}} \ , \tag{3.3}$$

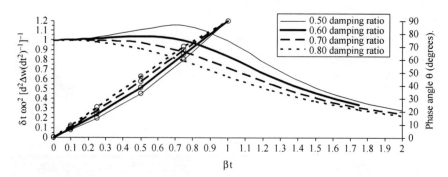

Fig. 3.10 Variation of the ratio between displacement δ_t of a SDOFO and its acceleration $d^2\Delta_w(dt^2)^{-1}$ normalized by ω_o^2 and change in phase angle θ of a SDOFO with change in the tuning ratio β_t

where δ_t is relative displacement of sensor, $\partial^2\Delta w\,(\partial t)^{-2}$ is ground acceleration, $\beta_t = f_d\,f_o^{-1}$ is the tuning ratio, f_d is the frequency of an input motion i.e. ground, f_o is the frequency of the output motion i.e. sensor, ξ is the damping ratio (Section 2.4.7). The phase angle θ between δ_t and $\partial^2\Delta w\,(\partial t)^{-2}$ is according to Equation (3.2). Figure 3.10 shows that a linear response to the excitation is best achieved for a damping ratio of about 0.6 and β_t smaller than 0.7. For this reason, the natural frequency of the sensor must be greater than the smallest recorded frequency.

Comments made in Section 3.2 for velocity transducers concerning instrument and system calibration, fixation and location are equivalent for accelerometers.

3.3.1 Analogue System

Such systems were mainly used in the past. Bormann (2002), for example, states for a completely analogue system, which contains sensors, signal conditioners, demultiplexers and analogue drum or film recorder, two primary drawbacks as:

- Low dynamic range and resolution of the acquired data causing their incompleteness. Many events have amplitudes that are too low to be resolved on paper or film records as well as that many records are clipped because their amplitude is too large for undistorted recording.
- Incompatibility of paper and film records with computer analysis.

3.3.2 Mixed Systems

Bormann (2002), for example, states that mixed systems have analogue sensors, analogue signal conditioning, analogue demultiplexers but digital data acquisition, digital processing and digital archiving. Such systems have a low dynamic range and, therefore, the same disadvantages as the analogue systems regarding data completeness and quality. However, they can accommodate off-line as well as automatic near-real time computer analysis. Most modern methods of data analysis can be used except those that require very high resolution raw data. Such systems can be useful when higher dynamic range of a fully digital system is not of prime importance. Advantages of these systems are low cost and low power consumption of the field equipment.

3.3.3 Digital Systems

In digital systems only sensors are analogue while all other components are digital. The dynamic range and the resolution are much higher than that of analogue and mixed type systems. Bormann (2002), for example, states that these factors depend mainly, but not only, on the number of bits of the analogue to digital converter. In practice, however, the total dynamic range and the resolution of data acquisition is usually less than the number of bits an analogue to digital converter would theoretically allow, since 24 bit converters rarely have a noise level as low as 1 bit.

Two known method exist (e.g. Bormann, 2002) that can further increase the dynamic range and/or the resolution of data recording.

- Gain ranging method automatically adjusts the analogue gain of the system according to the amplitude of the signal and thus prevents clipping of the strongest events. This significantly increases the dynamic range of data acquisition but with almost unchanged resolution. Unfortunately, even modern electronics are imperfect and gain ranging amplifiers introduce gain ranging errors in the data. Therefore, the resolution of gain ranged recording is actually decreased depending on the data, which makes these type of errors hard to detect. For this reason, many users are reluctant to use the gain ranging systems. They have been mostly replaced by straightforward, multi bit analogue to digital conversion, which allow nearly as wide a dynamic range.
- Over sampling principle is another approach, which helps to improve the dynamic range and resolution of digital acquisition. Data are samples at much higher rate than is required and then the value of each sample of the final (lower sampling rate) output data stream is calculated by a statistical model. The increase in the resolution is significant. However, the efficiency of over sampling depends on the ratio between the over sampling frequency and final sampling rate of actual data. The higher the final sampling rate used, the less benefit is gained for over sampling. Therefore, for a nearby event, which frequently requires 200 Hz sampled data, the benefit of over sampling is modest with some data logger designs.

An example of the use of a piezoelectric accelerometer for assessing vibration susceptibility over shallow and deep bedrock from weight drop sources follows.

3.3.3.1 Case Study of Assessed Vibration Susceptibility over Shallow and Deep Bedrock Using Accelerometers and Weight Drops

Paine (2003) states that *Measurements of ambient ground motion are adequate to characterize sites where noise sources already exist, but they inadequately characterize ground motion at undeveloped sites where noise sources (including laboratory activities) are not yet present.* A portable vibration source and measured induced ground motion as the source was placed at fixed distances from the sensor are used. The proposed site of new Metrology Laboratory in Austin, Texas, will be founded on/within deposits of sand, silt and clay over bedrock, which is deeper than 10 m below the ground surface. New laboratory is needed because a 1,130 kg mass movement by crane in the existing laboratory causes vertical acceleration greater than the 0.001 g threshold set for operating sensitive balances. Dominant ground motion frequencies during mass movement are a few tens of cycles per second. At another comparative site in Texas, limestone bedrock exists beneath clayey residual soil, which depth varies in the range from 23 to 178 cm.

To accurately measure small ground motion, which is nevertheless important for delicate instruments routinely used in the laboratory, a sensor that consists of three piezoelectric accelerometers fixed orthogonally in vertical and two horizontal directions was mounted on a machined steel block. The accelerometers have factory calibrated sensitivities of about 1,150 mV/g. Their voltage output is linear over acceleration range from 0.0001 g to about 5 g. The sensors measure acceleration to within 5% of the true magnitude at frequencies ranging from less than 0.1 to nearly 1000 Hz. Paine (2003) states that *In addition to their sensitivity to very small acceleration, accelerometers have other technical advantages over velocity based geophones.* The flat response to acceleration at a wide range of frequencies compared favourably with the decreased response of velocity based geophones at frequencies above and below their natural frequencies. Large signal magnitude over the entire useful vibration frequency range increases the achievable signal to noise ratio at the high frequency end of the spectrum, which also improves the subsurface imaging capabilities.

Measurements of ground acceleration at the proposed site reveal low levels of ambient ground motion. Peak acceleration is about an order of magnitude below the 0.001 g threshold. Records of background acceleration show no significant, coherent noise events. Some nearby activities affect the ground acceleration level. An irrigation pump operating 200 m away from the site increase ground acceleration slightly over background values. A tractor ploughing around the perimeter of the laboratory footprint further increased the peak accelerations but these peak values remain well below the acceleration threshold. Light traffic on a county road more than 200 m away produced peak and RMS ground accelerations that fell within the range recorded during background events.

Because seismic noise will be generated at the new laboratory that is not present there now, controlled active-source comparisons of the proposed site (deep bedrock) with the comparative shallow-bedrock site were conducted to examine the relative response of the sites to induced ground motion. The ground motion at the proposed and comparative sites were induced by dropping a 230 kg trailer mounted mass at 10 m intervals between distances of 10 and 100 m from the accelerometers. The recorded ground motion was analyzed by transferring the record to a computer, decoding to the raw voltage signal, converting the voltage to acceleration, integrating the acceleration to velocity and the velocity to displacement, and calculating peak and root mean square (RMS) values for voltage, acceleration, velocity and displacement.

At the proposed site, measured peak vertical accelerations reached 0.024 g and peak horizontal acceleration 0.077 g at a source to sensor distance of 10 m. At a source to sensor distance of 50 m, measured vertical acceleration is just below the 0.001 g threshold, however, horizontal accelerations exceeded the threshold value significantly and are dominated by slowly propagating, low frequency surface waves. Both vertical and horizontal accelerations remain below the 0.001 g threshold at a source to sensor distance of 100 m.

At the comparative site and when the source is 10 m from the sensor, vertical and horizontal accelerations exceeded the threshold value. Vertical accelerations are associated with high frequency direct and refracted longitudinal weaves, which amplitudes exceeded those generated during passage of lower frequency surface waves. Comparative weakness of the surface waves at 50 m from the source is also apparent in the horizontal acceleration record where the accelerations approach 0.001 g. At 100 m from the source, ground acceleration in all three directions remains below the 0.001 g threshold. Vertical accelerations are dominated by a 60 Hz signal caused either by ambient ground vibration from nearby electrical equipment or by induced electrical noise in the instrument or its connections. Lack of a dominant 60 Hz signal in the horizontal records suggests that the cause is ambient ground vibration.

The results of the measurements are shown in Fig. 3.11.

In summary:

- As expected, acceleration associated with induced ground motion declines rapidly with increasing distance from the source for all three components over both deep (proposed) and shallow (comparative) bedrock (site). The rate of peak acceleration decrease with distance is larger for the proposed site in comparison with the comparative site due to the wave radiation damping over half spherical wave fronts instead of circular ones respectively. On both sites, the peak horizontal accelerations exhibit smaller rates of decrease at distances greater than about 50 to 80 m than the vertical components.

- At 10 m distance from the source, peak accelerations are slightly higher at the proposed site (deep bedrock) than at the comparative site (shallow bedrock) in the vertical direction and are significantly higher over deep bedrock in the horizontal direction. Peak vertical accelerations are stronger than horizontal accelerations over shallow bedrock but are weaker than horizontal accelerations over deep bedrock.

- At source distances greater than 20 m, peak vertical accelerations are higher over shallow bedrock than over deep bedrock. At these distances, strong vertical

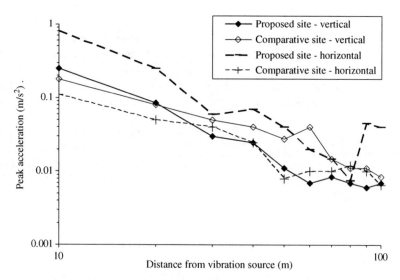

Fig. 3.11 Peak acceleration versus distance from vibration source in the case study in Section 3.3.3.1

ground motion associated with un-attenuated, high frequency body waves produces the largest acceleration over shallow bedrock while strong surface waves produce the highest vertical accelerations in soil layers overlying deep bedrock.

3.4 Summary

Ground vibration measurement is important for checking of amplitudes of predicted ground motion and for confirmation of efficiency of control measures of ground vibration. The type and locations of instruments used for vibration measurement depend not only on properties of a vibration source but also on the properties of measuring instrument.

- Velocity measurement transducers (sensors) termed geophones are used in the frequency range that is greater than the natural frequency of vibration of a velocity sensor, which is approximately from about 0.1 to 100 Hz in the case of so called short period sensors. Short period sensors are relatively stable in a broad range of temperatures and do not require power supply unless having electronically extended frequency range. Geophones are used at larger distances from high energy sources such as arising from pile driving and blasting to avoid decoupling (out of range) and aliasing (signal smoothing) effects.
- Acceleration measurement transducers (sensors) termed accelerometers are used in the frequency range that is smaller than the natural frequency of vibration of acceleration sensor, which is approximately a few tens of Hz unless being

electronically extended to a few kHz. Accelerometers require power supply and are more susceptible to background (environmental) noise than geophones, which are cheaper. Modern electronic miniature vibration sensors operate at a wider range of frequencies and amplitudes than conventional type accelerometers and geophones.

- Besides right choice of instrument type and their location, method of fixation of transducers to their bases is an important issue in ground vibration measurement. No slippage or separation of an instrument from its base must be allowed even in the case of very severe ground motion in the near field of blasting, which frequency could reach thousands of Hertz and amplitudes of several gravitational accelerations.

Chapter 4
Processing of Vibration Records

4.1 Introduction

Processing of vibration records is necessary because the visual inspection of a time history only reveals maximum amplitude and duration but not influences of potential noise caused by the recoding system/process and/or background (environment). Besides that, vibration records may contain various errors. Corrections of two basic errors are described in Sections 4.2 and 4.3. Douglas (2003), for example, listed types of possible non-basic errors in strong-motion records, Table 4.1: insufficient digitizer resolution, S-wave trigger, insufficient sampling rate, multiple baselines, spikes, early termination, and amplitude clipping.

These types of non-standard errors are shown in Fig. 4.1.

The objective of this chapter is to describe two types of basic errors in vibration records and the methods for their correction as well as spectral analyses of corrected records.

4.2 Filtering of High Frequencies

There are a number of reasons for filtering of high (and low) frequencies that may be contained within a vibration record. The following description is based on Bommer (1992) but not entirely.

- Vibration record may contain background noise together with the signal from vibration source, which vibration is intended to be monitored. Filters modify the recorded data to preserve information of importance but remove data where the noise to signal ratio is too high, usually at low and high frequencies within a vibration record. In filtering extreme parts of a record, both the noise and the true signal are lost.
- Low-pass (anti-aliasing) filter is used to ensure that, for all activities, the Nyquist frequency was well in excess of the frequencies anticipated, so that all wave forms were accurately recorded (e.g. Hiller and Crab, 2000). Aliasing is described in

M. Srbulov, *Ground Vibration Engineering*, Geotechnical, Geological, and Earthquake Engineering 12, DOI 10.1007/978-90-481-9082-9_4, © Springer Science+Business Media B.V. 2010

Table 4.1   Types of non-basic errors in strong motion records

Error type	Comments
Insufficient digitizer resolution	Digital instruments with a low bit range can cause recording of a few levels of amplitudes only resulting in a step like record. The resolution of the instrument can be calculated as $2A_f(2^{n_b})^{-1}$, where A_f is the instrument's amplitude range, n_b is the A/D converter bit range (usually between 10 and 24 bits). The peak ground acceleration (PGA) from instruments with low digitizer resolution is within $\pm r/2$ of the true PGA. The lack of sufficient bit range introduces high and low frequency noise into the recorded ground vibration.
S-wave trigger	Analogue instruments start recording once triggered by acceleration above a trigger level (say 0.001 g). Digital instruments record continuously but could lose the initial part of a record if their pre-event memory is too short.
Insufficient sampling rate	This error causes missing sections of data points.
Multiple baselines	Records from analogue instruments that were digitized in sections, which are not spliced together well.
Spikes	These can be corrected by either removing the suspect point(s) or reducing the amplitude of the spike after examining original record.
Early termination	Records from both analogue and digital instruments may exhibit lack of film or memory or instrument malfunction.
Amplitude clipping	Occurs when instrument measuring range is exceed by amplitudes of ground vibration.

Source: From Douglas (2003).

Section 4.2.1. Nyquist frequency is inversely proportional to twice the time interval at which vibration amplitudes are recorded. For example, for the analogue signal from a geophone sampled at approximately 2 kHz (0.0005 s time interval), a low-pass filtering is performed for frequencies above 800 Hz (e.g. Hiller and Crab, 2000).

High frequency noise is removed using filters. Analogue type filters within measuring instruments are no longer in use because they automatically filter useful information as well. Numerical filtering is used in frequency domain. For this reason, vibration records in time domain are transferred into frequency domain using fast Fourier transform (FFT), which is described in Section 4.2.1. The filter is applied by multiplication of the amplitudes of FFT and the filtered vibration record is recovered using the inverse transform. Alternatively, the impulse response of a filter can be found in time domain and applied to the vibration record in time domain by convolution.

An ideal numerical filter function would transmit required frequencies and attenuate completely all unwanted frequencies in a record of ground vibration. This, however, is not achievable due to limitations of the Fourier analysis. An instantaneous

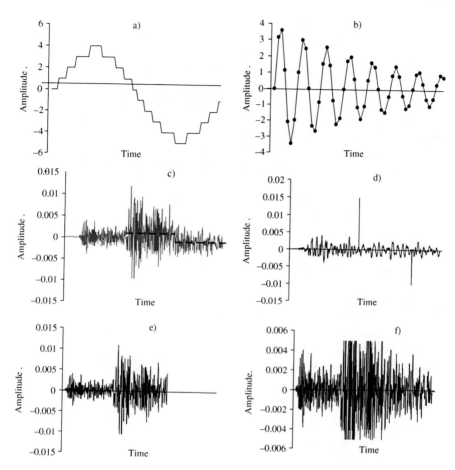

Fig. 4.1 Examples of vibration records caused by (**a**) insufficient instrument resolution, (**b**) insufficient sampling rate, (**c**) multiple base lines, (**d**) spikes, (**e**) early termination, (**f**) amplitude clipping

transition between wanted and unwanted frequencies requires having two simultaneous (step-like) values at the same frequency. This cannot be expressed by a Fourier series. To overcome this problem, a transition zone or ramp is introduced between narrow ranges of frequencies. Such a filter also is not perfectly realisable due to Gibb's effect (e.g. http://en.wikipedia.org/wiki/Gibbs_phenomenon) in which the Fourier representation of a function always overshoots the true value at a point of discontinuity. The effects of these limitations is to allow spectral leakage, thus Fourier amplitude spectrum (FAS), which is considered in Section 4.4.1, of a filtered signal divided by the FAS of the original signal would produce a rippled function. The extent of the ripples is largely dependent on the width of the transition zone. To some extent the ripples can be reduced by widening the transition zone. Various filters developed represent different attempts to overcome some of these difficulties and to achieve greater efficiency in terms of computing time. None of them is perfect.

The choice of the cut-off frequencies is made by the user performing the adjustment. A low frequency cut-off has a significant effect on the processed vibration record, since although of small amplitude, the duration of each wave can be large, and when integrated in time to obtain velocity and displacement can strongly influence their time histories. The high frequency cut-off has little influence since the waves are both of small amplitude and short duration. The high frequency limit is effectively controlled by the Nyquist frequency. It is inappropriate to apply filters to records of short duration. The reason is that for the filter to be effective it must modify the low frequency end of the FAS, with the first frequencies corresponding to $T_r^{-1}, 2T_r^{-1}$, etc., where T_r is the record duration. If a vibration record has duration of say 2s, then it means filtering out frequencies of 1/2, 2/2, etc. Hz, which represent periods of 0.5, 1, etc. seconds that may form part of the signal. Therefore, such filtering creates distortion of the record.

4.2.1 Fourier Analysis and Fast Fourier Transform

The analysis is described in many textbook such as by Chatfield (1992). Jean Baptise Joseph, baron de Fourier observed first in 1822 that any periodic function $f(t)$, with period T, can be represented by a series of sinusoidal and co sinusoidal waves (forming orthogonal pairs) with different phases and amplitudes, which are harmonically related. The Fourier series can be represented as (e.g. Bommer, 1992)

$$f(t) = \frac{a_0}{2} + \sum_{m=1}^{\infty} a_m \cdot \cos(m \cdot \omega_1 \cdot t) + \sum_{m=1}^{\infty} b_m \cdot \sin(m \cdot \omega_1 \cdot t)$$

$$= \frac{a_0}{2} + \sum_{m=1}^{\infty} c_m \cdot \sin(m \cdot \omega_1 \cdot t + \varphi_m), \tag{4.1}$$

$$where \quad c_m = \sqrt{a_m^2 + b_m^2}, \quad \varphi_m = \arctan\left(\frac{b_m}{a_m}\right), \quad \omega_1 = \frac{2 \cdot \pi}{T}$$

The series $f(t)$ is applicable to a function, which is repeated at intervals of T_f from $-\infty$ to $+\infty$. The coefficients in Equation (4.1) are:

$$a_0 = \frac{2}{T_f} \cdot \int_{-T_f/2}^{T_f/2} f(t) \cdot dt$$

$$a_m = \frac{2}{T_f} \cdot \int_{-T_f/2}^{T_f/2} f(t) \cdot \cos(m \cdot \omega_1 \cdot t) \cdot dt \quad m = 1, 2, ..., \infty \tag{4.2}$$

$$b_m = \frac{2}{T_f} \cdot \int_{-T_f/2}^{T_f/2} f(t) \cdot \sin(m \cdot \omega_1 \cdot t) \cdot dt \quad m = 1, 2, ..., \infty$$

The Fourier amplitude spectrum (FAS) is the graphical representation of the values of c_m for each frequency $m\omega_1$, which is the frequency-domain representation of the time function. For a periodic function, FAS will actually be a series of discrete values, each corresponding to the frequency of each of the infinite number of sinusoidal functions that together represent the time function $f(t)$. The FAS can also be determined for non-periodic functions, using the Fourier transform (or Fourier integral):

$$F(\omega) = \int_{-\infty}^{+\infty} f(t) \cdot [\cos(\omega \cdot t) + i \cdot \sin(\omega \cdot t)] \cdot dt = \int_{-\infty}^{+\infty} f(t) \cdot \exp(-i \cdot \omega \cdot t)\, dt$$

(4.3)

In this case, ω dos not correspond to a set of discrete values but to a continuous function. The FAS of periodic and non-periodic functions are shown in Fig. 4.2. From Fig. 4.2, it follows that continuous functions in time have non-periodic spectra and vice versa.

$F(\omega)$ is a complex quantity:

$$F(\omega) = |F(\omega)| \cdot \exp[i \cdot \varphi(\omega)] = F_R(\omega) + i \cdot F_I(\omega),$$

(4.4)

where $F_R(\omega)$ and $F_I(\omega)$ are the real and imaginary parts of the FAS respectively, $\varphi(\omega)$ is the Fourier phase spectrum (FPS). Hence,

$$|F(\omega)| = \sqrt{F_R{}^2(\omega) + F_I{}^2(\omega)}, \quad \varphi(\omega) = \arctan\frac{F_I(\omega)}{F_R(\omega)}$$

(4.5)

If the complex function $F(\omega)$ is known, then the original time history can be recovered using the inverse Fourier transform:

$$f(t) = \frac{1}{2 \cdot \pi} \cdot \int_{-\infty}^{+\infty} F(\omega) \cdot \exp(i \cdot \omega \cdot t) \cdot dw$$

(4.6)

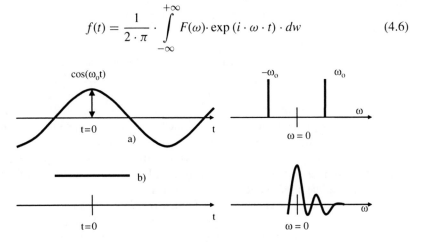

Fig. 4.2 Example Fourier amplitude spectra (FAS) of (**a**) periodic, (**b**) non-periodic functions

When Fourier analysis is applied to discrete in time vibration records, they are considered not as continuous functions, which original analogue signal is, but as a series of discrete points, at equal time intervals Δt. The vibration amplitude $a(t)$ is considered to be represented by $a[n]$, $n = 1, \ldots, N$ and time t is expressed as $t = (n-1)\Delta t$. The discrete Fourier transform is:

$$F_d(\omega) = \sum_{n=1}^{N} a[n] \cdot \exp\{-i \cdot \omega \cdot (n-1) \cdot \Delta t\} \cdot \Delta t \tag{4.7}$$

$F_d(\omega)$ is actually only determined at a number of discrete frequencies, the smallest is inversely proportional to the duration of the record $(N-1)\Delta t$, and the other is integer multiples:

$$\omega_k = \frac{2 \cdot \pi \cdot k}{T_f} = \frac{2 \cdot \pi \cdot k}{(n-1) \cdot \Delta t} \qquad k = -\frac{N}{2}, \ldots, \frac{+N}{2}, \tag{4.8}$$

where N is assumed to be an even number. The DFT can be defined as:

$$F[k] = \sum_{n=1}^{N} a[n] \cdot \exp\left\{\frac{-2 \cdot \pi \cdot i \cdot (n-1)}{N-1}\right\} \cdot \Delta t \tag{4.9}$$

The inverse is:

$$a[n] = \frac{1}{2 \cdot \pi} \cdot \sum_{k=-N/2}^{+N/2} F[k] \cdot \exp\left\{\frac{2 \cdot \pi \cdot i \cdot (n-1)}{N-1}\right\} \cdot \Delta\omega \tag{4.10}$$

Although $F(\omega)$ is a continuous function for an non-periodic signal, the FAS that is obtained from the DFT is a series of discrete points, which closely represent the continuous spectrum. The frequencies at which the FAS is known are separated by T_f^{-1}. If T_f is infinite then the spacing will be zero and the spectrum will be continuous.

Application of DFT can cause problem called aliasing. This occurs because the transformation of a discrete function in time domain into frequency domain produces a periodic function, i.e. $F[k]$ is repeated every Δt^{-1} Hertz i.e. $2\pi \Delta t^{-1}$ radians/s in terms of the circular frequency. If a continuous function is defined within $\pm\omega$ range, sampling at Δt causes this spectra to be repeated every $2\pi \Delta t^{-1}$ radians/s. If ω is greater than $\pi \Delta t^{-1}$ (i.e. if the highest frequency in the spectrum ω is greater than $(2\Delta t)^{-1}$, the Nyquist frequency) then there will be overlapping and frequencies above the Nyquist frequency will be distorted lower frequencies. This is aliasing ('folding') and it needs to be considered when selecting the sampling rate and interpreting the high frequency part of the FAS.

Because DFT is only known over a limited range of frequencies, in the range between T_f^{-1} and $(2\Delta\tau)^{-1}$ (the Nyquist frequency in order to avoid problem of aliasing) the record cannot be fully recovered using the inverse transform, since this requires integration over the frequency range from $-\infty$ to ∞, i.e. o to ∞ for real records. The transform from time to frequency is valid since although the integral is also between ±∞, the transient vibration record is zero at times less than 0 and greater than T_f, so that no information is lost.

Calculation of DFT is rather inefficient. Instead, an algorithm that is called the fast Fourier transform (FFT) after Cooley and Tukey (1965) is used. The implementation requires that the vibration record consist of N samples, where $N = 2^q$, q is an integer. For a selected sampling interval Δt, q needs to be chosen so that $(N-1)\Delta t$ is long as the duration of the vibration record, which may then require the addition of zeros beyond the end of the vibration record. The original sequence, $a[n]$, is then divided into two subsets representing terms in odd and even positions, $x[n] = a[2n]$, $y[n] = a[2n-1]$, $n = 1, \ldots, N/2$. The DFT of each short sequence is calculated from

$$X[k] = \sum_{n=1}^{N/2} x[n] \cdot \exp\left\{\frac{-2 \cdot \pi \cdot i \cdot k \cdot (2n)}{N-1}\right\} \cdot \Delta t \qquad k = 1, \ldots, N$$

$$Y[k] = \sum_{n=1}^{N/2} y[n] \cdot \exp\left\{\frac{-2 \cdot \pi \cdot k \cdot (2 \cdot n - 1)}{N-1}\right\} \cdot \Delta t \quad k = 1, \ldots, N \qquad (4.11)$$

and

$$F[k] = X[k] + Y[k] \cdot \exp\left\{\frac{-i \cdot 2 \cdot \pi \cdot k}{N-1}\right\} \qquad k = 1, \ldots, N$$

The transforms $X[k]$ and $Y[k]$ can each be calculated from the transforms of two further subsets of $N/4$ points, and since N is an integer power of 2, the process can be continued until the Σ terms contain only one term.

4.2.1.1 Example of Fast Fourier Transform and Filtering in Frequency Domain

The example uses a built in Microsoft Excel Tools/Data Analysis/Fourier analysis procedure capable of handling up to 2048 record points as shown in Section 1 of Appendix. Longer records can be separated into peaces each containing not more than 2048 record points. The record in time domain is shown in Fig. 4.3.

Amplitudes of a discrete Fourier series (with 2048 members) in the positive side of the frequency range (because of the logarithmic scale used, which does not allow negative numbers) obtained by the built in fast Fourier transform procedure within Microsoft Excel are shown in Fig. 4.4 from Section 1 of Appendix.

The Nyquist frequency of the record is $(2 \times 0.005)^{-1} = 100$ Hz and the corresponding circular frequency is 2π times larger i.e. 628 radians/s, which is exactly the highest frequency present in the record. If someone is not interested in high frequencies then a filtering of the Fourier amplitudes is applied. In the example, the filtering is performed using chosen function $\{\exp[-(\omega_n - 300)/100]^2\}^{-1}$ for the angular frequencies greater than 300 radians/s, as shown in Fig. 4.5 together with the Fourier amplitudes of the filtered record from Fig. 4.3.

Using inverse FFT in Section 1 of Appendix, the filtered record in time domain is shown in Fig. 4.6.

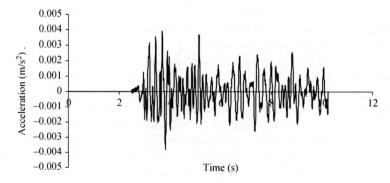

Fig. 4.3 Acceleration time history considered in the example in Section 4.2.1.1

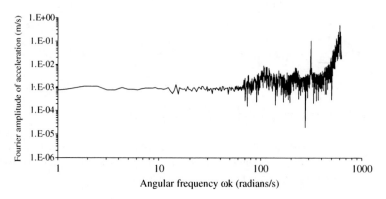

Fig. 4.4 Amplitudes of a discrete Fourier series as a representation in frequency domain of the time series in Fig. 4.3

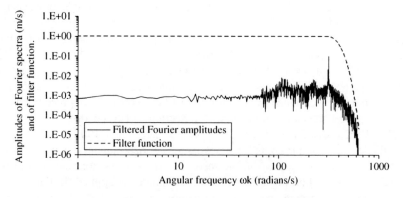

Fig. 4.5 Amplitudes of filter function and filtered Fourier series from Fig. 4.3

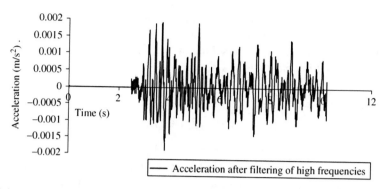

Fig. 4.6 Acceleration time history of filtered record from Fig. 4.5

From Fig. 4.6, it can be seen that the filtering of high frequencies caused a substantial decrease of the peak acceleration. It means that the peak acceleration amplitudes occur with high frequency pulses containing little energy and of little significance for structural damage. Sarma and Srbulov (1998) showed, using a large number of acceleration records, that 95% of the vibration energy described by a record is realized within acceleration level of about $\pm 2/3$ of the peak acceleration. Therefore 2/3 of the peak acceleration is more representative value than the peak value.

4.3 Baseline Correction

It is inappropriate to apply filters to records of short duration as explained in Section 4.2, last paragraph. The baseline correction consists of finding a function that approximates the long period offset from the time axis (baseline error) of a vibration record. The error could be caused by digitization of an analogue record or other reasons. If uncorrected, the base line error may cause erroneous values of ground velocity and displacement when they are integrated in time from an acceleration record.

The best fit of the record average values (baseline) is then subtracted from the record considered to produce desirable adjustment. Different criteria may be used to define the best fit. Sometimes, the least square approach is used i.e. minimization of the sum of squared differences between uncorrected and corrected record. In other cases it may be required to achieve zero end velocity. The baseline correction can provide reasonable results for the records of short duration. Depending on the shape of the record average values, different functions can be considered: from linear to polynomial of higher order and other functions. For example, when a polynomial of 4th order is considered for a baseline error function its expression is:

$$a(t) = C_1 \cdot t^4 + C_2 \cdot t^3 + C_3 \cdot t^2 + C_4 \cdot t + C_5, \tag{4.12}$$

where $a(t)$ is the average value of the record at time t. To determine unknown coefficients C_{1-5}, it is necessary to use, in this case, five values $a(t)_{1-5}$ of the averaging function at five times t_{1-5}. The resulting system of five linear equations with five unknowns in matrix notation is:

$$
\begin{bmatrix} a_1(t) \\ a_2(t) \\ a_3(t) \\ a_4(t) \\ a_5(t) \end{bmatrix} = \begin{bmatrix} t_1^4 & t_1^3 & t_1^2 & t_1 & 1 \\ t_2^4 & t_2^3 & t_2^2 & t_2 & 1 \\ t_3^4 & t_3^3 & t_3^2 & t_3 & 1 \\ t_4^4 & t_4^3 & t_4^2 & t_4 & 1 \\ t_5^4 & t_5^3 & t_5^2 & t_5 & 1 \end{bmatrix} \times \begin{bmatrix} C_1 \\ C_2 \\ C_3 \\ C_4 \\ C_5 \end{bmatrix} \tag{4.13}
$$

The unknown constants C_{1-5} are obtained from the following matrix form:

$$
\begin{bmatrix} C_1 \\ C_2 \\ C_3 \\ C_4 \\ C_5 \end{bmatrix} = \begin{bmatrix} t_1^4 & t_1^3 & t_1^2 & t_1 & 1 \\ t_2^4 & t_2^3 & t_2^2 & t_2 & 1 \\ t_3^4 & t_3^3 & t_3^2 & t_3 & 1 \\ t_4^4 & t_4^3 & t_4^2 & t_4 & 1 \\ t_5^4 & t_5^3 & t_5^2 & t_5 & 1 \end{bmatrix}^{-1} \times \begin{bmatrix} a_1(t) \\ a_2(t) \\ a_3(t) \\ a_4(t) \\ a_5(t) \end{bmatrix} \tag{4.14}
$$

where the superscript $^{-1}$ denotes matrix inversion and $\times$ matrix multiplication. The calculation can be performed using a Microsoft Excel spreadsheet, which contains predefined functions MINVERS(array) and MMULT(array1, array2) for inversion and multiplications of matrices respectively, as shown in Section 2 of Appendix.

4.3.1 Example of Baseline Correction for the Record Shown in Fig. 4.3

The averaging function of the record shown in Fig. 4.3 but with multiple baselines is obtained using Microsoft Excel/Data/Add Trend line/Polynomial of the 4th order as shown in Fig. 4.7 from Section 2 of Appendix.

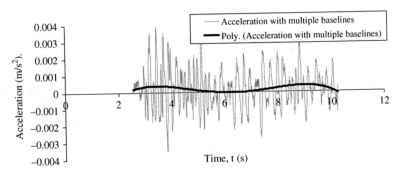

Fig. 4.7 Fourth order polynomial fitted by Microsoft Excel to the acceleration record from Fig. 4.3 but with multiple baselines in the example in Section 4.3.1

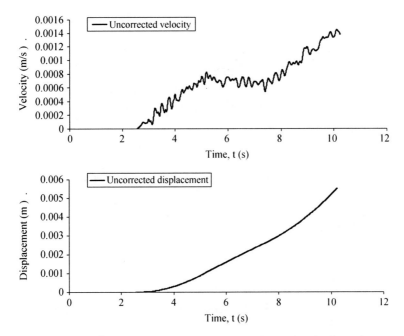

Fig. 4.8 Velocity and displacement records obtained by integration in time of uncorrected acceleration record from Fig. 4.7

The velocity and displacement records obtained by integration in time from the uncorrected acceleration record show their departures from the time axis as shown in Fig. 4.8.

The coefficients of the 4th order polynomial have been determined using Equation (4.14) as shown in Section 2 of Appendix. The baseline function is:

$$a(t) = -0.000000076 \cdot t^4 + 0.000188 \cdot t^3 - 0.00162 \cdot t^2 + 0.0056 \cdot t - 0.00628 \quad (4.15)$$

The resultant baseline corrected acceleration record, after subtraction of the baseline function, is shown in Fig. 4.9 from Section 2 of Appendix.

It is possible that even after baseline correction of an acceleration record, the velocity and displacement obtained by integration in time of the corrected record still require baseline corrections of themselves as shown in Fig. 4.10.

The baseline correction of velocity and displacement records is similar to the baseline correction of the acceleration record. First the displacement record is corrected using the polynomial from Section 2 of Appendix.

$$d(t) = -0.000001011 \cdot t^4 + 0.000034513 \cdot t^3 - 0.0003028 \cdot t^2 + 0.00096351 \quad (4.16)$$

The baseline corrected displacement record is shown in Fig. 4.11.

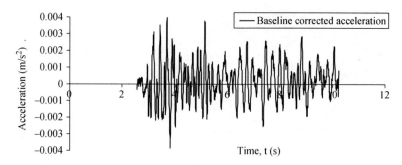

Fig. 4.9 Baseline corrected record from Fig. 4.3 but with multiple baselines

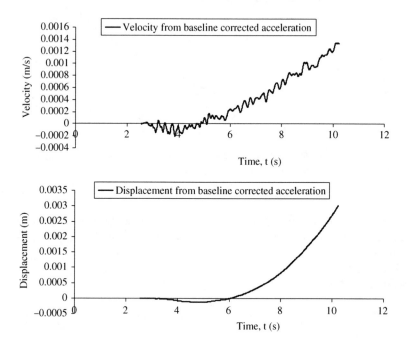

Fig. 4.10 Velocity and displacement obtained by integration in time of corrected acceleration record from Fig. 4.9

Velocity is first derivative of the corresponding displacement, so that the first derivative of the displacement baseline from Equation (4.16) needs to be subtracted from the velocity record in Fig. 4.10. The resulting corrected velocity record is shown in Fig. 4.12. The end velocity is not zero because not complete but only a part of an actual record is considered in the example.

Acceleration is first derivative of the corresponding velocity, so that the second derivative of the displacement baseline from Equation (4.16) needs to be subtracted from the acceleration record shown in Fig. 4.9. The resulting corrected record is shown in Fig. 4.13.

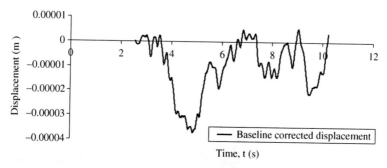

Fig. 4.11 Baseline corrected displacement record from Fig. 4.10

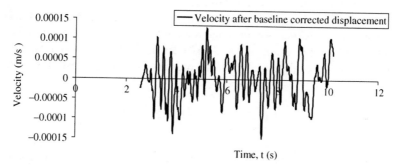

Fig. 4.12 Velocity record from Fig. 4.10 after subtraction of the first derivative of the displacement baseline

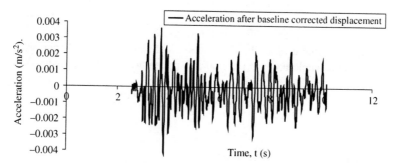

Fig. 4.13 Final corrected acceleration record from Fig. 4.9 after subtraction of the second derivative of the displacement baseline

4.4 Spectral Analyses

Single peak amplitude of a vibration record is a poor representation of whole time history and therefore a whole range (spectrum) of the amplitudes of a record with respect to the frequency (or period) is frequently considered. The following spectra are considered:

- Fourier spectra
- Power spectra
- Response spectra

4.4.1 Fourier Spectra

A plot of Fourier amplitudes versus frequency (c_m versus $m\omega_1$ in Equation 4.1) is called a Fourier amplitude spectrum (FAS). Similarly, a plot of Fourier phase angle (ϕ_m versus $m\omega_1$ in Equation 4.1) is called Fourier phase spectrum, which exhibits the variation of a ground motion with time.

As stated by Bommer (1992), among others, FAS (described in Section 4.2.1) has many applications, particularly in geophysics, and is a useful tool in the analysis of ground vibrations. For example, the area under the FAS is directly related to the ground motion energy described by the record according to the Parseval's relation:

$$\int_o^t a^2(t) \cdot dt = \frac{1}{\pi} \cdot \int_0^\infty F_A^2(\omega) \cdot d\omega \qquad (4.17)$$

where t is time, $a(t)$ is the amplitude of ground vibration record in time domain, $F_A(\omega) = c_m(\omega)$ is the amplitude of Fourier transform in frequency domain, ω is circular frequency $\omega = 2\pi f$. Considerable information about the vibration source parameters can be obtained from FAS. The choice of type of spectra is based on the frequency range considered: displacement spectra are used for examination of the low frequency content while acceleration spectra are used for examination of high frequency content of a vibration record. The amplitudes of displacement (D), velocity (V) and acceleration (A) of Fourier spectra are directly related:

$$F_v(\omega) = \frac{F_A(\omega)}{\omega}, \quad F_D(\omega) = \frac{F_A(\omega)}{\omega^2} \qquad (4.18)$$

4.4.1.1 Example Shapes of FAS

Example shapes of FAS of acceleration (A) and displacement (D) after smoothing are shown in Fig. 4.14.

The values of the corner frequency f_c and the amplitude Ω_o are directly related to the properties of a vibration source. The general shape of the far field displacement spectra will consist of a horizontal part at low frequencies and an inclined part. The reason for this is that the displacement signal for any component of either longitudinal or transversal wave should be a unidirectional pulse, without oscillations, the spectrum of this pulse will have its maximum value at zero frequency (infinite period), which corresponds to the static case and is the maximum displacement at the source. For this reason, the low frequency trend in the displacement spectrum

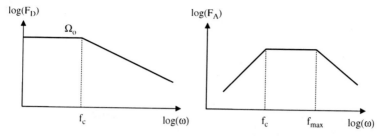

Fig. 4.14 Example shapes of Fourier amplitude spectra (FAS) of acceleration (*A*) and displacement (*D*)

should be a line parallel to the frequency axis. The amplitude of this low frequency pulse is proportional to the source energy. The corner frequency f_c, which defines the high frequency limit, is related to the dimensions of the energy source. For a circular source shape, f_c is proportional to the ratio of the transversal wave velocity and the radius of the source. The cut-off frequency f_{max} is a property of local site condition, but may include also source dependency. A narrow range between f_c and f_{max} implies that the motion has a dominant frequency (or period), which is characteristic of a smooth almost sinusoidal time history. This can be typical for the motions in far fields when the high frequencies have been filtered due to material damping. Consequently, the shapes of the spectra are also distance dependent. Also, soil tends to exhibit the peak values at smaller frequencies in comparison with rocks. Such a difference is less obvious from the time histories of vibrations in comparison with the spectral values.

4.4.2 Power Spectra

The frequency content of a ground motion can also be described by a power spectrum or power spectral density function (e.g. Vanmarke, 1976) as

$$S(\omega) = \frac{1}{\pi \cdot T} \cdot \left| \int_0^T a(t) \cdot \exp(-i \cdot \omega \cdot t) \cdot dt \right|^2 = \frac{F_A^2(\omega)}{\pi \cdot T} \qquad (4.19)$$

The power spectrum shows how the intensity of ground accelerations or energy per unit time (power) at a given point in space is distributed over the frequencies. One of the properties of the power spectrum is that its integral in frequency domain is a measure of root mean square (rms) value of the ground acceleration i.e.

$$a^2{}_{rms} = \int_0^\infty S(\omega) \cdot d\omega \qquad (4.20)$$

As in the case of definition of root mean square amplitude, the problem with this definition is the choice of period T.

4.4.3 Response Spectra

Typical response spectra, which are used extensively in earthquake engineering, are graphs showing maximum response of an elastic single degree of freedom oscillator (SDOFO) on a rigid base (Fig. 4.15) to the ground vibration time history as a function of the natural period (or frequency) of vibration of the oscillator for a given amount of damping.

Not all structures can be approximated by a SDOFO. Single-storey building frame, bridge with hinged columns and multi-storey structure responding in rigid base condition are typical examples of structures that can be represented as a SDOFO. Examples of non SDOFO are arch bridges, shells and domes, irregular elevated frames, etc.

The main assumptions of this model are:

- The first and only possible vibration mode is the most significant.
- Soil-structure interaction effect is not significant and is not considered.
- Horizontal and vertical ground motions caused by earthquakes are considered separately.

The SDOFO model has frequently been considered for generation of response spectra, since their introduction by Benioff (1934) and Biot (1941). Variants of the basic SDOFO are also considered in earthquake engineering. For example, the effect of soil-structure interaction is considered approximately by using rotational and translational soil springs as shown in Fig. 4.16.

Recent examination of different elastic SDOFOs for horizontal structural response to combined horizontal and vertical ground motion caused by earthquakes was done by Ambraseys and Douglas (2003). Their main findings are:

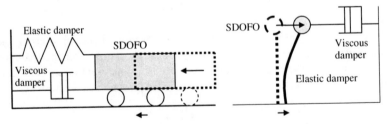

Fig. 4.15 Single degree of freedom oscillator (SDOFO) on a rigid base with models of a box like and a beam like structure with elastic (spring/beam) and viscous (dash pot) dampers

Fig. 4.16 Linear elastic
SDOFO on a flexible base
(soil) without viscous
damping

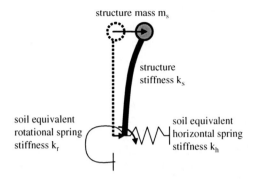

- Bending and hinging SDOFO has three main types of behaviour: normal, parametric resonance and instability.
- The type of behaviour that a system exhibits is controlled by the combination of system parameters and the vertical input acceleration. As a consequence of variation of input acceleration, a system can exhibit all three types of behaviours.

Linear elastic response spectra are appropriate for linear elastic structural force-displacement response. Strong ground motion may induce inelastic behaviour as shown by an idealized force-displacement function in Fig. 4.17.

The inelastic behaviour is frequently characterized using ductility factor $\mu = d_{max} (d_y)^{-1}$, where d_y is the yield displacement, d_{max} is the maximum allowable displacement. The inelastic response spectra can be calculated directly for defined μ, for example Ambraseys et al (2004). For example, Eurocode 8-1 paragraph 3.2.2.5(2) states *To avoid explicit inelastic structural analysis in design, the capacity of the structure to dissipate energy, through mainly ductile behaviour of its elements and/or other mechanisms, is taken into account by performing an elastic analysis based on a response spectrum reduced with respect to the elastic one, henceforth called a 'design spectrum'. This reduction is accomplished by introducing the behaviour factor q.* Other design codes use similarly called reduction factors of elastic response spectra of SDOFO on a rigid base.

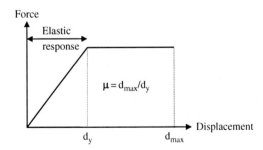

Fig. 4.17 Idealized inelastic
force-displacement function

The most commonly used method for determination of linear elastic response spectra nowadays is direct solution of the governing equation of motion of SDOFO on rigid base, when the time history of ground motion is treated as a series of linear accelerations (e.g. Bommer, 1992). The equation of motion of a SDOFO on a rigid base is:

$$\ddot{u} + 2 \cdot \xi \cdot \omega_o \cdot \dot{u} + \omega_o{}^2 \cdot u = -\left[A_{i-1} + B_i \left(t_i - t_{i-1} \right) \right], \qquad (4.21)$$

where u is relative displacement of a SDOFO with respect to the rigid base, $\dot{u}$ and $\ddot{u}$ are the first and second derivative in time i.e. relative velocity and acceleration of a SDOFO, $\omega_o = k_o{}^{1/2} m_o{}^{-1/2}$, k_o is the stiffness (force per displacement) of an elastic damper shown in Fig. 4.15 i.e. a structure and soil equivalent spring stiffness shown in Fig. 4.16, m is the mass of a SDOFO i.e. of a structure and its foundation, damping ratio $\xi = c_o (2 m_o \omega_o)^{-1}$, c is an equivalent viscous damper (Fig. 4.15) parameter (force times time per length), A_{i-1} is base acceleration at time t_{i-1}, B_i is the gradient of an acceleration record in time interval $\delta t_i = t_i - t_{i-1}$. The solution of Equation (4.21) is found in two parts as usual; the first for a homogenous case with the right hand side equal to zero and then finding the particular integral corresponding to the right hand side function to obtain at each time instant i for a usually under-dumped SDOFO (critically and over dumped SDOFO do not experience oscillations) and:

$$u_i = \exp\left(-\xi \cdot \omega \cdot \delta t_i\right) \cdot \left[C_i \cdot \sin\left(\omega_d \cdot \delta t_i\right) + D_i \cdot \cos\left(\omega_d \cdot \delta t_i\right) \right] - \frac{B_i}{\omega_o{}^2} \cdot \delta t_i$$

$$- \frac{A_{i-1} - 2 \cdot \xi \cdot B_i/_{\omega_o}}{\omega_o{}^2}$$

$$\dot{u}_i = \exp\left(-\xi \cdot \omega_o \cdot \delta t_i\right) \cdot \left[-\left(C_i \cdot \xi \cdot \omega_o + D_i \cdot \omega_d \right) \cdot \sin\left(\omega_d \cdot \delta t_i\right) \right.$$

$$\left. + \left(C_i \cdot \omega_d - D_i \cdot \xi \cdot \omega_o \right) \cdot \cos\left(\omega_d \cdot \delta t_i\right) \right] - \frac{B_i}{\omega_o{}^2}$$

$$\ddot{u}_{abs} = \ddot{u}_i + a_{base} = -2 \cdot \xi \cdot \omega_o \cdot \dot{u}_i + \omega_o{}^2 \cdot u_i$$

$$C_i = \frac{\dot{u}_{i-1} + D_i \cdot \xi \cdot \omega_o + B_i/_{\omega_o{}^2}}{\omega_d}$$

$$D_i = u_{i-1} + \frac{A_{i-1} - 2 \cdot \xi \cdot B_i/_{\omega_o}}{\omega_o{}^2},$$

$$(4.22)$$

where the circular frequency of a dumped oscillator $\omega_d = \omega_o (1 - \xi)^{1/2}$. For usual $\xi \le 0.2$ it follows that $\omega_d \sim \omega_o$. Elastic displacement response spectrum can be obtained from related elastic acceleration response spectral values by their division with ω^2 (e.g. Eurocode 8, Part 1, 2004).

Pseudo spectra (i.e. pseudo relative velocity and the pseudo absolute acceleration spectra determined from the relative displacement spectrum by multiplying of its ordinates by circular frequency and the squared circular frequency respectively) were originally introduced for convenience when the calculation of spectral ordinates was time consuming. Pseudo-spectra can be determined by direct integration in time of the equation of forced vibration of a SDOFO. For values of $\xi \leq 0.2$, pseudo spectral acceleration is almost identical to the spectral acceleration; however, pseudo spectral velocity may be quite different from the spectral velocity, as large as 20% depending on the frequency and damping. Despite that, pseudo spectra are still widely used, perhaps because of habits. Determination of structural frequency and damping is considered in Section 5.3.

4.4.3.1 Example of an Elastic Acceleration Response Spectra

Acceleration, velocity and displacement elastic response spectra, which are shown in 4.18 for the record from Fig. 4.3, are obtained using Section 3 of Appendix.

Figure 4.18 shows that the peak spectral acceleration, velocity and displacement occur at different frequencies (periods). For this reason, response spectra are called acceleration controlled (at high frequency), velocity controlled (at intermediate frequency) and displacement controlled (at low frequency) portions. Also, it is evident that the peak values are strongly dependent on damping. Most structural codes specify that 5% damping is applicable during strong earthquakes. However, damping depends on many factors, such as:

- Structural material (steel, concrete, timber, brick)
- Structural type (frame, plate, shell, panel, masonry)
- Types of joints (fixed, hinged, frictional)
- Intensity of ground vibration (small vibration causes small deformation and small damping and vice versa)

Determination of fundamental period of oscillation (frequency) of a future foundation (and structure above it) is described in the following Sections 5.3 and 5.4. Fundamental period (frequency) and damping of an existing structure is best determined from filed tests of vibration of structures (e.g. Hall, 1987).

4.5 Summary

Processing of vibration records is necessary because the visual inspection of a time history only reveals maximum amplitude and duration but not influence of potential noise in a record caused by the recoding system/process and/or background (environment). Besides that, vibration records may contain various errors. Non-basic errors can be determined by visual inspection of a record as shown in Fig. 4.1. These errors are (e.g. Douglas, 2003):

- Insufficient digitizer resolution
- S-wave trigger
- Insufficient sampling rate
- Multiple baselines
- Spikes

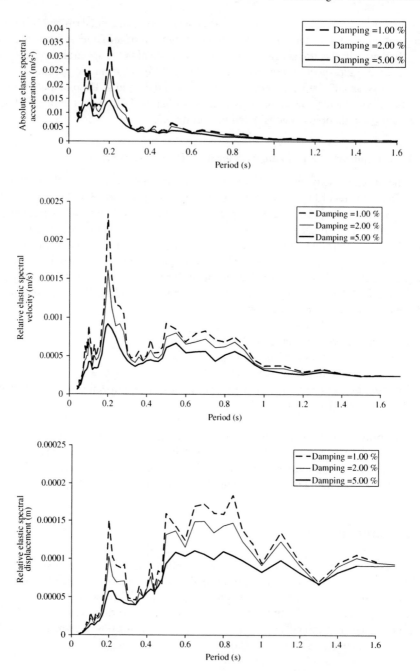

Fig. 4.18 Elastic acceleration, velocity and displacement response spectra for the record shown in Fig. 4.3

- Early termination
- Amplitude clipping

Basic errors such as baseline offset and inappropriate frequency content can be corrected using routine techniques. There are a number of reasons for filtering of high (and low) frequencies that may be contained within a vibration record.

- Vibration record may contain background noise together with the signal from vibration source, which vibration is intended to be monitored.
- Low-pass filter is used to ensure that, for all activities, the Nyquist frequency was well in excess of the frequencies anticipated, so that all wave forms were accurately recorded (e.g. Hiller and Crab, 2000). Nyquist frequency is inversely proportional to twice the time interval at which vibration amplitudes are recorded.

High frequency noise is removed using filters. For this reason, vibration records in time domain are transferred into frequency domain using fast Fourier transform (FFT), which is described in Section 4.2.1. The filter is applied by multiplication of the amplitudes of FFT and the filtered vibration record is recovered using the inverse transform.

It is inappropriate to apply filters to records of short duration. Baseline correction consists of finding a function that approximates the long period offset from the time axis (baseline error) of a vibration record. The error could be caused by digitization of an analogue record or other reasons. If uncorrected, the base line error may cause erroneous values of ground velocity and displacement when they are integrated in time from an acceleration record.

Single peak amplitude of a vibration record is rather poor representation of whole time history and therefore a whole range (spectrum) of the amplitudes of a record with respect to the frequency (or period) is frequently considered. The following spectra are considered:

- Fourier spectra
- Power spectra
- Response spectra

A plot of Fourier amplitudes versus frequency is called a Fourier amplitude spectrum (FAS). Similarly, a plot of Fourier phase angle is called Fourier phase spectrum, which exhibits the variation of a ground motion with time. Typical response spectra, which are used extensively in engineering, are graphs showing maximum response of an elastic single degree of freedom oscillator (SDOFO) on a rigid base to the ground vibration time history as a function of the natural period (or frequency) of vibration of the oscillator for a given amount of damping. Determination of fundamental period of oscillation (frequency) of a future foundation (and structure above it) is described in the following Sections 5.3 and 5.4. Fundamental period (frequency) and damping of an existing structure is best determined from filed tests using vibration actuators placed on the structure.

Chapter 5
Foundation and Structure Effects

5.1 Introduction

Foundation (and structure above) can either amplify or attenuate amplitudes of incoming ground waves depending on a number of parameters that are described later. Two types of soil-foundation (and structure) interactions are commonly referred to in the literature.

- 'Kinematic' interaction is caused by inability of a foundation to follow ground motion due to greater foundation stiffness in comparison with ground stiffness. In effect, stiff foundation filters high frequency ground motion to an averaged translational and rotational foundation motion. Average values are smaller than the maximum values and therefore 'kinematic' interaction is beneficial except if averaged motion results in significant rotation and rocking of a foundation.
- 'Inertial' interaction is caused by the existence of structural and foundation masses. Seismic energy transferred into a structure is dissipated by material damping and radiated back into ground causing superposition of incoming and outgoing ground waves. As a result, the ground motion around a foundation can be attenuated or amplified, depending on a variety of factors. The most important factor in determining the response is the ratio between the fundamental period of a foundation and the fundamental period of adjacent ground in the free field. The ratio of unity indicates resonance condition between foundation and its adjacent ground, which is to be avoided.

The objective of this chapter is to describe a simplified model for consideration of soil-foundation kinematic interaction and for determination of fundamental frequency/period of a simplified soil and foundation model for consideration of inertial interaction effects based on response spectra, which are described in Section 4.4.3.

5.2 A Simplified Model of Kinematic Soil-Foundation Interaction

Due to differences in ground and shallow footing/pile stiffness, averaging of ground motion over footing/pile length is termed kinematic soil-foundation interaction. Newmark et al. (1977) proposed a simple procedure for averaging of free-field

M. Srbulov, *Ground Vibration Engineering*, Geotechnical, Geological, and Earthquake Engineering 12, DOI 10.1007/978-90-481-9082-9_5, © Springer Science+Business Media B.V. 2010

ground motion. Sarma and Srbulov (1996) used this approach for analysis of a number of case histories including piled and shallow foundations. Although the inclination of incoming ground waves is frequently near vertical at shallow depths near vibration sources, the spatial wave incoherence occurs as a result of ground heterogeneity. Also, shallow foundations are frequently affected by the propagation of near surface waves (Raylegh and Love). Such averaged (filtered) ground acceleration is then used for the analyses of inertial interaction i.e. for response spectra.

The expression for the average acceleration a_t over the length of a footing/pile at time t (e.g. Sarma and Srbulov, 1996) is

$$
\begin{aligned}
a_t &= \frac{1}{L_s} \int_{L-L_s}^{L} a_l dl \\
&= \frac{c_t}{L_s} \int_{t-T_s}^{t} a_t dt \\
&= \frac{1}{T_s}(v_{tp} - v_{t-T_s}) ,
\end{aligned}
\tag{5.1}
$$

where L_s is the length along footing/pile over which ground motion is averaged, L is the distance from footing/pile beginning to a referent point along the surface/depth, a_l is ground acceleration at length/depth l at time t, c_t is soil transversal waves velocity and is assumed equal to the velocity of transversal waves passing through ground along the footing/pile, T_s is the time (in seconds) necessary for a seismic wave to pass along L_s and must be less or equal to the ratio between the footing/pile length and c_t, v_{tp} and v_{t-T_s} are the ground velocities at the footing/pile beginning at times t and $t-T_s$. The ground velocities can be obtained from corresponding acceleration time history by numerical integration in time.

5.2.1 Example of the Kinematic Soil-Foundation Interaction Effect

The acceleration time history shown in Fig. 4.3 is used for the example. It is assumed that a shallow strip footing under rigid panels has length of 20 m and that soil transversal waves velocity under the footing is 200 m/s so that $T_s = 20/200 = 0.1$ s. Microsoft Excel workbook of Section 3 of Appendix is used to obtain averaged footing acceleration shown in Fig. 5.1 and the response spectra shown in Fig. 5.2.

From Figs. 4.3 and 5.1, it can be seen that the footing peak acceleration is about a half in comparison with the free field ground acceleration.

From a comparison of Figs. 4.18 and 5.2 it can be seen that the peak spectral amplitudes of a structure resting on the shallow strip footing in the example are about 2/3 of the spectral amplitudes of structures resting on isolated pads. The differences in the spectral amplitudes are negligible for periods greater than

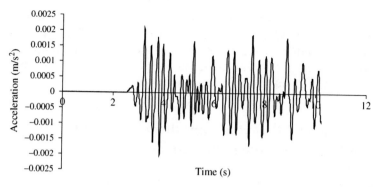

Fig. 5.1 Averaged acceleration of the shallow footing in the example in Section 5.2.1

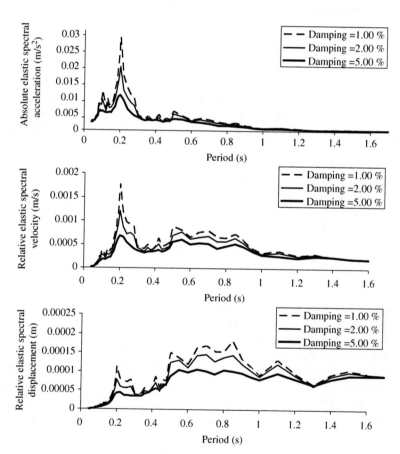

Fig. 5.2 Elastic spectral values for a structure resting on the shallow footing in the example in Section 5.2.1

about 0.3 s. Figure 5.2 also shows that the peak acceleration of foundation and structures with the fundamental vibration period of about 0.2 s will be amplified due to resonance effect by a factor of 10 for damping of about 1% with respect to the peak acceleration of a mass-less foundation shown in Fig. 5.1. Assessment of foundation and structural vibration period is described in the following section.

5.3 Fundamental Period of Vibration of a Simplified Soil-Foundation Interaction Model

The fundamental circular frequency $\omega_e = 2\pi T_e^{-1}$ of coupled linear elastic single degree of freedom oscillator (SDOFO) without viscous damping (Fig. 4.16) is calculated from the following formula (e.g. Wolf, 1994):

$$\frac{1}{\omega_e^2} = \frac{1}{\omega_s^2} + \frac{1}{\omega_h^2} + \frac{1}{\omega_r^2} , \tag{5.2}$$

where ω_s is the natural circular frequency of foundation and structure in fixed base condition, ω_h is equal to the natural frequency of the dynamic model in the horizontal direction, assuming that piles are rigid (infinitely stiff) and that the foundation cannot rock (the rocking stiffness is infinite), ω_r equals the natural frequency corresponding to the rocking motion for pile groups (infinitely stiff) and with no horizontal motion of the foundation (the horizontal stiffness is infinite). In reality, the horizontal and rocking motions of foundations are coupled and not independent. The natural circular frequency ω_s of foundation and structure for rigid base condition is:

$$\omega_s = \sqrt{\frac{k_s}{m_s}} , \tag{5.3}$$

where k_s is the coefficient of stiffness of foundation and structure, m_s is their mass.

Determination of the natural circular frequencies of foundation in horizontal ω_h and rocking ω_r motion can be rather complex. Novak and Grigg (1976), Poulos (1979), Wolf (1994), among others, used the concept of dynamic interaction factors between only two piles within a pile group. The presence of other piles is disregarded; the corresponding reflections and refractions are not taken into account. In the simplified method used in this section, shallow foundation is represented by an equivalent disk. A deep foundation is modelled by a stack of embedded disks over the foundation depth in the strength-of-material approach by Wolf and Deeks (2004).

The circular frequencies ω_h and ω_r are determined from the peaks of ratios between the amplitudes of the horizontal and rocking motion of equivalent embedded disks and the amplitudes of free-field ground motion. The free-field ground motion for vertically propagating shear waves with circular frequency ω is described by Wolf and Deeks (2004) as

$$u^f(z, \omega) = u^f(\omega) \cdot \cos \frac{\omega}{c_t} \cdot z, \tag{5.4}$$

where the depth z is measured downwards from the free surface, $u^f(\omega)$ is the surface amplitude of the free field ground motion, ω is the ground circular frequency, c_t is ground transversal wave velocity. Computer program CONAN (by Wolf and Deeks, 2004, http://w3.civil.uwa.edu.au/~deeks/conan/) is used for the calculations. Layered soil sites and half space or rock bases can be considered. Soil properties required for the calculations are unit density ρ, Poison's ratio v, shear modulus G and damping ratio ξ_g. The radii of the equivalent disks for the vertical motion r_v are calculated as:

$$r_v = r_{pile} \cdot \sqrt{N_{piles}}$$

$$r_v = \sqrt{\frac{b_c \cdot l_c}{\pi}} \tag{5.5}$$

The radii r_h of the disks for the horizontal motion are

$$r_h = r_{pile} \cdot \sqrt[4]{N_{piles}}$$

$$r_h = \sqrt[4]{\frac{b_c \cdot l_c^3}{3 \cdot \pi}} \text{ or } \sqrt[4]{\frac{b_c^3 \cdot l_c}{3 \cdot \pi}}, \tag{5.6}$$

depending if a raft foundation/pile cap rotate around its shorter (b_c) or longer (l_c) side respectively. The radii r_r of the disks for the rotational motion are

$$r_r = \sqrt[4]{\frac{4 \cdot I_r}{\pi}}$$

$$I_r = \sum_1^{N\ piles} \left(\frac{r_{pile}^4 \cdot \pi}{4} + r_{pile}^2 \cdot \pi \cdot y_{pile}^2 \right), \tag{5.7}$$

$$I_r = \frac{b_c \cdot l_c^3}{12} \text{ or } \frac{b_c^3 \cdot l_c}{12}$$

where r_{pile} is a half of pile diameter, y_{pile} is the shortest distance between pile centroid and the neutral axis of rotation, b_c and l_c are the breadth and length of a rectangular foundation. The rocking motion is negligible for relatively thin flexible foundations.

The effects of material damping and of radiation damping are separated (Wolf, 1994) to derive simple expressions that lead to physical insight. This is achieved by considering the effect of material damping on the damping coefficients only. The equivalent hysteretic damping ratio ξ_e determined at resonance is used over the whole range of frequency of a coupled SDOFO (Wolf, 1994)

$$\xi_e = \frac{\omega_e^2}{\omega_s^2} \cdot \xi_s + \left(1 - \frac{\omega_e^2}{\omega_s^2} \right) \cdot \xi_g + \frac{\omega_e^2}{\omega_h^2} \cdot \xi_h + \frac{\omega_e^2}{\omega_r^2} \cdot \xi_r, \tag{5.8}$$

where ξ_s is structural hysteretic damping ratio, ξ_g is soil hysteretic damping ratio, radiation damping ratio in horizontal direction is ξ_h and in rotational motion of a pile group is ξ_r. The values of ω_e and ω_s are given in Equations (5.2) and (5.3) respectively. The circular frequencies ω_h and ω_r are determined from the peaks of ratios between the amplitudes of the horizontal and rocking motion of equivalent disks and the amplitudes of free-field ground motion using the computer program CONAN mentioned earlier.

Clough and Penzien (1993) described the procedure for determination of ξ_s; ξ_g can be obtained from Table 4.1 of Eurocode 8–5, for example, which for the peak horizontal accelerations of 0.1, 0.2 and 0.3 of the gravitational accelera- tion shows ξ_g of 0.03, 0.06 and 0.1 respectively. Srbulov (2008) indicates ξ_g of 0.125 for the peak horizontal acceleration of 0.5 of the gravitational acceleration. Table 4.1 of Eurocode 8–5 also gives the ratio between shear modulus G and its maximum value $G_{max} = \rho c_t^2$, where ρ is ground unit density and c_t transversal wave velocity, depending on the peak ground acceleration. Srbulov (2008) indi- cates $G\,G_{max}^{-1}$ of 0.20 ± 0.15 for the peak horizontal acceleration of 0.5 of the gravitational acceleration.

The radiation damping ratio in the horizontal direction ξ_h of a foundation is according to Wolf (1994)

$$\xi_h = \frac{a_h \cdot z_h}{2 \cdot r_h}$$

$$z_h = \pi \cdot r_h \cdot (2 - v)/8, \tag{5.9}$$

$$a_h = \frac{\omega_h \cdot r_h}{c_t}$$

where r_h is according to Equation (5.6), v is soil Poisson's ratio, c_t is soil transversal wave velocity, ω_h is the circular frequency of horizontal motion.

The radiation damping ratio in rotational motion ξ_r of a foundation is according to Wolf (1994)

$$\xi_r = \frac{a_r \cdot c_r}{2 \cdot k_r} \tag{5.10}$$

For $v < 1/3$,

$$c_r = \frac{z_r \cdot c_t}{r_r \cdot c_p}$$

$$z_r = \frac{9}{8}\pi(1 - v) \cdot r_r$$

$$a_r = \frac{\omega_r \cdot r_r}{c_p} \tag{5.11}$$

$$k_r = 1$$

and for $1/3 < v < \frac{1}{2}$

$$c_r = \frac{z_r}{2 \cdot r_r}$$

$$z_r = \frac{9}{8}\pi(1-v) \cdot r_r$$

$$k_r = 1 - 0.6 \cdot \left(v - \frac{1}{3}\right) \cdot \frac{z_r}{r_r} \cdot a_r^2,$$

$$a_r = \frac{\omega_r \cdot r_r}{2 \cdot c_t}$$

(5.12)

where r_r is according to Equation (5.7), v is soil Poisson's ratio, c_t is soil transversal wave velocity, c_p is soil longitudinal wave velocity, ω_r is the circular frequency in rotational motion.

The circular frequency of a dumped oscillator $\omega_d = \omega_e(1-\xi_e)^{1/2}$. For usual $\xi_e \geq 0.2$ it follows that $\omega_d \sim \omega_e$.

5.3.1 Generalized Single Degree of Freedom Oscillator

Clough and Penzien (1993) provide expressions for generalized SDOFO shown in Fig. 5.3 as a beam with variable mass m_x along its length, with a concentrated mass m_i, and mass moment of inertia $I_{m,i}$ at a place i, with axial force N_x, with distributed soil reactions k_x, and a concentrated spring k_j at a place j. The beam deflected position is described by an assumed function Ψ_x and the end displacement Z_t in time.

The parameters in Equation (5.3) are:

$$m_s = \int_0^{L_b} m_x \cdot \psi_x^2 \cdot dx + \sum m_i \cdot (\psi_i)^2 + \sum I_{m,i} \left(\frac{d\psi_x}{dx}\right)_i^2$$

$$k_s = \int_0^{L_b} k_x \cdot \psi_x^2 \cdot dx + \int_0^{L_b} E \cdot I_x \cdot \left(\frac{d^2\psi_x}{dx^2}\right)^2 \cdot dx$$

(5.13)

$$+ \sum k_j \cdot (\psi_j)^2 - \int_0^{L_b} N_x \cdot \left(\frac{d\psi_x}{dx}\right)^2 \cdot dx,$$

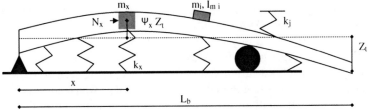

Fig. 5.3 Generalized SDOFO parameters

where E is Young modulus of the beam material and I_x is the second moment of cross section area of the beam.

5.3.2 Case Study of Determination of the Fundamental Frequency of Vibration of a Caisson

Kaino and Kikuchi (1988) described the response of a bridge pier supported by a caisson type foundation during an earthquake in Japan and their analyses of the response by finite element method. The magnitude 6.7 earthquake that occurred East of Chiba Prefecture on 17 December 1987 caused the peak horizontal ground acceleration of about 0.03 of the gravitational acceleration.

The caisson (embedded concrete cylinder) is 10 m by 12 m wide in plan with 1.5 m thick walls 35.7 m long, with the top placed 5.5 m below the ground surface so its depth is 41.2 m. The bridge pier is 6 m by 10.5 m wide with 1.5 m thick walls 19.65 m high. From Equations (5.6) and (5.7) it follows that the radii of the equivalent embedded stacked cylinders for the horizontal and rocking motion are 5.21 m for the pier and 6.54 m for the caisson. The Fourier spectra of the recorded horizontal accelerations at the bridge pier top exhibits the peak value at 1.6 Hz, at the caisson top at 0.8 Hz and the caisson bottom at 1.8 Hz with a strong peak value at 0.8 Hz.

Soil profile at the location of the caisson consists mainly of very loose silty sand and sandy silt to a depth of 22.5 m, clay and medium dense sand to a depth of 40 m and very dense sandy gravel and silt to a depth of 50 m. Distribution of transversal wave velocity with depth is given in Table 5.1.

Table 5.1 Transversal wave velocity versus depth in the case study in Section 5.3.2

Depth (m)	0–5	5–25	25–30	30–34	34–37	37–45
c_t (m/s)	80	140	200	260	210	350

Source: From Kaino and Kikuchi (1988).

Assumed data for further analyses are: Poisson's ratio of 0.35 to a depth of 25 m and 0.25 at greater depths, damping 1%, unit density of 1700 kg/m^3 to a depth of 25 m and 1900 kg/m^3 at greater depths. The ratios between caisson and ground vibration amplitudes in the horizontal direction are shown in Fig. 5.4 and for rocking in Fig. 5.5 based on the CONAN results.

The fundamental period of vibration in the horizontal direction of the caisson is about 2 Hz according to the first crest in Fig. 5.4 and in rocking about 1.8 Hz according to the first crest in Fig. 5.5 at frequencies greater than zero. Combined vibration frequency according to Equation (5.2) is 1.3 Hz, which corresponds to the frequency of the first mode of caisson vibration that is indicated in the paper by Kaino and Kikuchi (1988).

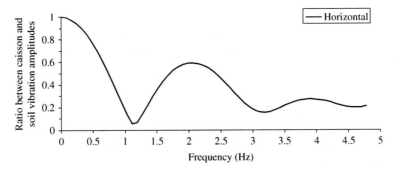

Fig. 5.4 Ratios between caisson and ground vibration amplitudes in the horizontal direction in the case study in Section 5.3.2

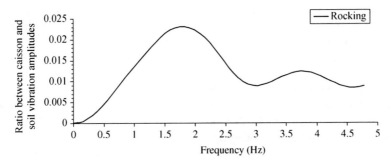

Fig. 5.5 Ratios between caisson and ground vibration amplitudes during rocking in the case study in Section 5.3.2

5.3.3 Case Study of Determination of the Fundamental Frequency of Vibration of Foundation of a Large Scale Shaking Table

Tajimi (1984) described the results of measurements and analyses of the fundamental periods of vibration of the foundation of a large scale shaking table.

The foundation is 44.8 m wide and 90.9 m long in plan with varying depth from 21 m at the centre to 13 m at the edge, Fig. 5.6.

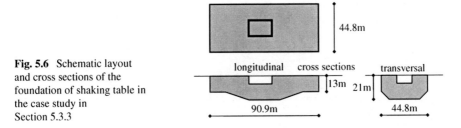

Fig. 5.6 Schematic layout and cross sections of the foundation of shaking table in the case study in Section 5.3.3

Table 5.2 Transversal wave velocity versus depth in the example

Depth (m)	0–8	8–13	13–21	21–48	48–78	78–117	117–181	>117
c_t (m/s)	160	200	320	360	400	520	640	1160

Source: From Tajimi (1984).

From Equations (5.6) and (5.7) it follows that the radii of the equivalent embedded stacked cylinders for the longitudinal and rocking motion are 43.5 m to a depth of 13 and 18.1 m at the bottom. The Fourier amplitude spectra of measured free vibration of the foundation exhibit the peak value at about 0.8 Hz in the longitudinal direction and 2.0 Hz in rocking.

Ground profile at the location of the foundation contains gravel and sand to a depth of 48 m, silty clay to a depth of 78 m, sand and silt to a depth of 117 m, gravel to a depth of 181 m and weathered granite at greater depths. Distribution of transversal wave velocity with depth is given in Table 5.2.

Assumed data for further analyses are: Poisson's ratio of 0.35 to a depth of 13 m and 0.25 at greater depths, damping 1%, unit density of 1700 kg/m^3 to a depth of 13 m and 1900 kg/m^3 at greater depths. The ratios between caisson and ground vibration amplitudes in the horizontal direction are shown in Fig. 5.7 and for rocking in Fig. 5.8 based on the CONAN results.

Figure 5.7 indicates the fundamental period of foundation vibration in the horizontal direction of about 2.5 Hz according to the first convex point on the graph at frequencies larger than zero. The 2.5 Hz frequency coincides with the period of horizontal vibration of the ground at a foundation depth of 21 m. The first crest in Fig. 5.8 at frequencies larger than zero indicates the fundamental frequency of vibration in rocking at about 2.5 Hz. Combined vibration frequency of the horizontal and rocking vibration according to Equation (5.2) is 1.8 Hz, which is close to a local peak at 2.0 Hz observed in the Fourier amplitude spectra of recorded vibrations in the longitudinal direction.

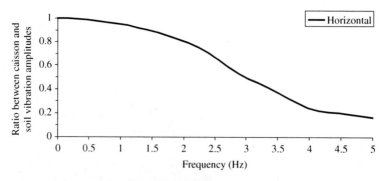

Fig. 5.7 Ratios between foundation and ground vibration amplitudes in the horizontal direction in the case study in Section 5.3.3

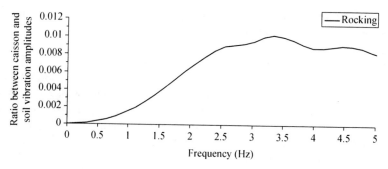

Fig. 5.8 Ratios between foundation and ground vibration amplitudes during rocking in the case study in Section 5.3.3

5.3.4 Case Study of Determination of the Fundamental Frequency of Vibration of a Seven-Story Reinforced Concrete Building in Van Nuys–California

Ivanovic et al. (2000) describe the results of two detailed ambient vibration surveys of a 7-story (6 floors above the ground level) reinforced concrete building in Van Nuys, California. Both surveys were conducted after the building was severely damaged by the 17 January 1994 Northridge earthquake, with $M_L = 6.4$ at the epicentral distance of 1.5 km, and its aftershocks. The first survey was conducted on 4 and 5 February 1994; the second one on 19 and 20 April 1994, about one month after the 20 March aftershock, with $M_L = 5.3$ at the epicentral distance of 1.2 km. The vibration frequencies of the building and two and three dimensional modal shapes for longitudinal, transversal and vertical vibrations were calculated. The surveys were not successful in identifying the highly localized damage (Fig. 5.9) by simple spectral analyses of the ambient noise data. Ivanovic et al. (2000) suggested that high

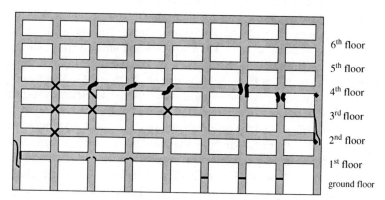

Fig. 5.9 Schematic locations of damaged columns (cracks from 0.5 to ~5 cm wide) in frames D (*thinner lines*) and A (*thicker lines* – larger cracks) of the building in the case study in Section 5.3.4

spatial resolution of recording points is required to identify localized column and beam damage, which is identified by a visual inspection.

The building is 19.1 m wide, 45.75 m long and 20.0 m high. The first floor is located 4.1 m above the ground floor; the other floors are spaced at 2.65 m. The reinforced concrete frames have the columns at 6.1 m centres in the transversal direction and at 5.8 m in the longitudinal direction, 38 columns in total. Spandrel beams surround the perimeter of the structure. Lateral forces in each direction are resisted by interior column-slab frames and exterior column spandrel beam frames. The structure is essentially symmetrical with the exception of some light framing members supporting the staircase and elevator openings. The plan configurations of each floor, including the roof, are the same except for two small areas at the ground floor. The floors are reinforced concrete flat slabs, 0.25 m thick at the first floor above ground, 0.215 m thick at the second to sixth floors above ground and 0.20 m thick at the roof. The north side of the building has four bays of brick masonry walls located between the ground and the floor above at the East end of the structure. Nominal 25 mm expansion joints separate the walls from the underside of the spandrel beams of the first floor above ground. The building is located on Holocene (recent) alluvium, with thickness up to 30 m and with the average shear wave velocity of 300 m/s over the 30 m depth. The building is supported by groups of two to four cast in place reinforced concrete piles of 0.61 m diameter and about 12.2 m length. The piles are located under the columns and the pile caps are connected by a grid of beams.

For both experiments, the measurements were taken along the middle longitudinal frame at nine columns on each floor and in three directions of motion. Three seismometers were used to measure the building response and three seismometers were used to measure the motions at the reference site on the ground level. For each experiment, two calibration tests were conducted, at the beginning and end of the experiment. These tests consisted of bringing all the six instruments close to each other and simultaneously recording the outputs. The purpose of these tests was comparison of the recorded amplitudes. The duration of each of the recordings during testing and survey was about 3 minutes and the sampling rate was set to 400 readings per second. The transducers were placed either directly on concrete or ceramic tiles. The experiments were carried continuously throughout day. During the first experiment, strong wind of about 50 km/h was blowing intermittently. The temperature was in the range from 8 to 15°C. It was a week day and typical heavy traffic was moving at 100 to 200 m from the building site. At the roof of the building, the air conditioning equipment was working continuously. The elevators were not in use. Before the second experiment, the building was restrained by wooden braces to increase the structural capacity near the areas of structural damage after a strong aftershock with epicentre at 1.2 km from the building. The braces were placed on the first three stories above ground floor at selected spans of the external frames. Only the ground floor of the interior longitudinal frames was restrained. There were no braces added to the transverse frames. During the second experiment, the temperature during the sunny days was in a range from 12 to 25°C; no elevators or air-conditioning were in operation.

Fourier spectra were computed from each record by FFT (following amplitude correction based on the calibration test results) and the transfer function of the vibration amplitudes were computed for each measuring point and component of motion with respect to the appropriate reference point on the ground floor. The Fourier amplitude spectra were smoothed while the transfer function amplitudes were not smoothed. The phase angles for the transfer functions were computed. In drawing the apparent shape functions, the phase angle was approximated by 0 or by π. No vertical modes of vibration could be identified from the transfer function of recorded vertical motions. The first experiment results indicated the system frequencies in the longitudinal direction of 1.0, 3.5, 5.7 and 8.1 Hz for the first four modes of vibration. For the transverse direction, the identified system frequencies were 1.4, 1.6, 3.9 and 4.9 Hz for the first four modes of vibration. The frequencies of 1.4 and 3.9 Hz correspond to translational modes and 1.6 and 4.9 Hz correspond to torsional vibration modes. The second experiment results indicated the system frequencies in the longitudinal direction of 1.1, 3.7, 5.7 and 8.5 Hz and in the transversal direction of 1.4, 1.6, 4.2 and 4.9 Hz for the first four nodes of vibration. For the frequency of 8.5 Hz, the signal to noise ratio was small, and hence it was difficult to analyze the phases.

It is interesting to compare the frequencies obtained based on the results of the experiments with the frequencies calculated using the simplified methods that are described in this chapter. Dimensions of the columns are scaled from the drawing in the paper by Ivanovic et al. (2000). The scaling resulted in the widths of 38 columns in the longitudinal direction of the building of 0.43 m (17") and in the transversal direction 0.85 m (33.5"). Other values used in Equation (5.13) are as follow:

- $I_x = 38*0.85*0.43^{3}*12^{-1} = 0.214$ m^4 in the longitudinal and $38*0.43*0.85^{3}*12^{-1} = 0.836$ m^4 in the transversal direction of the building
- E is assumed $= 20$ GPa for reinforced concrete in long term condition
- $m_x = 38*0.43*0.85*2500 = 34722.5$ kg/m
- k_x and $k_j = 0$ as the columns are not resting on external springs (cables/struts)
- The concentrated masses m_i and mass moment of inertia $I_{m,i} = 0.0833*m_i*$ (L or T length)^2 (e.g. Gieck and Gieck, 1997) at a place i (floors above ground and roof) in the longitudinal (L) or transversal (T) direction of the building are given in Table 5.3.

Table 5.3 Masses and mass moments of inertia of the floors and roof in the case study in Section 5.3.4

Floor	m_i (kg)	$I_{m,i}$ (kg$\times$m^2)	
		Longitudinal	Transversal
First above ground	540422	94261419	16087012
Others	464763	81064834	13834833
Roof	432337	75409030	12869592

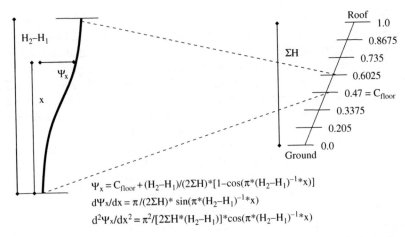

$$\Psi_x = C_{floor} + (H_2{-}H_1)/(2\Sigma H)*[1{-}\cos(\pi*(H_2{-}H_1)^{-1}*x)]$$
$$d\Psi_x/dx = \pi/(2\Sigma H)* \sin(\pi*(H_2{-}H_1)^{-1}*x)$$
$$d^2\Psi_x/dx^2 = \pi^2/[2\Sigma H*(H_2{-}H_1)]*\cos(\pi*(H_2{-}H_1)^{-1}*x)$$

Fig. 5.10 Assumed local shape function Ψ_x and its derivatives and global shape function coefficients C_{floor} in the case study in Section 5.3.4

The calculation is performed per floor using an assumed local and global shape functions with the coefficients C_{floor} shown in Fig. 5.10 for the first mode of vibration. The local function is assumed based on a deformed shape of a beam with non-rotating ends with one end displaced horizontally because of much greater flexural rigidity of the floor slabs in comparison with the flexural rigidity of the columns. The global shape is adopted as a simple stack of the local shapes for primarily the horizontal motion of the building.

Details of the calculation of the generalized mass m_s from Section 5.3.1 are given in Table 5.4 where

Table 5.4 Generalized mass m_s calculation in the case study in Section 5.3.4

Floor No.	Height H_i above ground (m)	$\int_0^{L_b} m_x \cdot [\Psi_x]^2 dx$	$m_i \cdot (\Psi_i)^2$	$I_{m,i} \cdot \left(\dfrac{d\Psi_x}{dx}\right)_i^2$
	20.0	–	432337	0
7	17.35	214125	349760	0
6	14.7	−54079	251076	0
5	12.05	134364	168712	0
4	9.4	−46006	102666	0
3	6.75	67527	52933	0
2	4.1	−25010	22711	0
1	0	2244	–	–
	Sum	293165	1380195	0

$$\int_0^{L_b} m_x \cdot \Psi_x^2 dx = m_x \cdot \sum_1^7 \left| \int_{H_1}^{H_2} \left\{ C_{floor} + \frac{H_2 - H_1}{2 \cdot \Sigma H} \cdot \left[1 - \cos\left(\frac{\pi}{H_2 - H_1} \cdot x \right) \right] \right\}^2 dx \right|$$

$$= \sum_1^7 \left| m_x \cdot \left(C_{floor} + \frac{H_2 - H_1}{2 \cdot \Sigma H} \right)^2 \cdot (H_2 - H_1) - 2 \cdot m_x \cdot \left(C_{floor} + \frac{H_2 - H_1}{2 \cdot \Sigma H} \right) \right.$$

$$\left. \cdot \frac{(H_2 - H_1)^2}{2 \cdot \Sigma H \cdot \pi} \cdot \left[\sin\left(\frac{\pi}{H_2 - H_1} \cdot H_2 \right) - \sin\left(\frac{\pi}{H_2 - H_1} \cdot H_1 \right) \right] + \left(\frac{H_2 - H_1}{2 \cdot \Sigma H} \right)^2 \cdot m_x \right.$$

$$\left. \cdot \left\{ \frac{H_2 - H_1}{2} + \frac{H_2 - H_1}{4 \cdot \pi} \cdot \left[\sin\left(\frac{2 \cdot \pi}{H_2 - H_1} \cdot H_2 \right) - \sin\left(\frac{2 \cdot \pi}{H_2 - H_1} \cdot H_1 \right) \right] \right\} \right| \tag{5.14}$$

The $m_s = 293165 + 1380195 = 1673360$ kg. A simple summation of all the masses would provide a total mass of 3991025 kg (+138%). Details of the calculation of the generalized stiffness k_s from Section 5.3.1 are given in Table 5.5 where

$$\int_0^{L_b} E \cdot I_x \cdot \left(\frac{d^2 \Psi_x}{dx^2} \right)^2 dx = \sum_1^7 \left| E \cdot I_x \cdot \int_{H_1}^{H_2} \left[\frac{\pi^2}{2 \cdot \Sigma H \cdot (H_2 - H_1)} \cdot \cos\left(\frac{\pi}{H_2 - H_1} \cdot x \right) \right]^2 dx \right|$$

$$= \sum_1^7 \left| E \cdot I_x \cdot \frac{\pi^4}{[2 \cdot \Sigma H \cdot (H_2 - H_1)]^2} \right.$$

$$\left. \cdot \left\{ \frac{H_2 - H_1}{2} + \frac{H_2 - H_1}{4 \cdot \pi} \cdot \left[\sin\left(\frac{2 \cdot \pi}{H_2 - H_1} \cdot H_2 \right) - \sin\left(\frac{2 \cdot \pi}{H_2 - H_1} \cdot H_1 \right) \right] \right\} \right| \tag{5.15}$$

Table 5.5 Generalized stiffness k_s calculation in the case study in Section 5.3.4

No	Height H_i above ground (m)	$\int_0^{L_b} E \cdot I_x \cdot \left(\frac{d^2 \Psi_x}{dx^2} \right)^2 dx$		$-\int_0^{L_b} N_x \cdot \left(\frac{d\Psi_x}{dx} \right)^2 dx$
		Longitudinal	Transversal	
	20.0			
7	17.35	49164	192061	−143
6	14.7	49164	192061	−296
5	12.05	49164	192061	−450
4	9.4	49164	192061	−604
3	6.75	49164	192061	−757
2	4.1	49164	192061	−911
1	0	31777	124137	−1685
	Sum from 0 to 20 m	326761	1276505	−4846

$$-\int_0^{L_b} N_x \cdot \left(\frac{d\Psi_x}{dx}\right)^2 dx = \sum_1^7 !- \left|\int_{H_1}^{H_2} N_x \cdot \left[\frac{\pi}{2 \cdot \Sigma H} \cdot \sin\left(\frac{\pi}{H_2 - H_1} \cdot x\right)\right]^2 dx\right|$$

$$= \sum_1^7 \left|-\frac{N_x \cdot \pi^2}{(2 \cdot \Sigma H)^2} \cdot \left\{\frac{H_2 - H_1}{2} - \frac{H_2 - H_1}{4 \cdot \pi} \cdot \left[\sin\left(\frac{2 \cdot \pi}{H_2 - H_1} \cdot H_2\right) - \sin\left(\frac{2 \cdot \pi}{H_2 - H_1} \cdot H_1\right)\right]\right\}\right|$$

$$(5.16)$$

The $k_s = 326761-4846 = 321915$ kN/m in the longitudinal and $1276505-4846 = 1271659$ kN/m in the transversal direction. A simple summation of all the stiffness of the columns would provide a total stiffness of $12\,E\,I_x\,l^{-3}$, which is the ratio between the reaction force and applied horizontal displacement at the end of a beam with prevented rotation at its ends (e.g. Jenkins, 1989). For the longitudinal direction, the stiffness would be 745201 kN/m (+131%) and in the transversal direction 2911159 kN/m (+129%) when $l = 4.1$ m on the ground floor.

The fundamental frequency of vibration in the longitudinal direction of the building on a rigid base is according to Equation (5.3) $(2\pi)^{-1}$ $(321915^*10^3{}^*1673360^{-1})^{1/2} = 2.2$ Hz and in the transversal direction $(2\pi)^{-1}$ $(1271659^*10^3{}^*1673360^{-1})^{1/2} = 4.4$ Hz. If lumped mass and simple stiffness of a SDOFO shown in Fig. 4.15 are used then the fundamental frequency of vibration in the horizontal direction of the building on a rigid base is according to Equation (5.3) $(2\pi)^{-1}$ $(745201^*10^3{}^*3991025^{-1})^{1/2} = 2.2$ Hz and in the transversal direction $(2\pi)^{-1}$ $(2911159^*10^3{}^*3991025^{-1})^{1/2} = 4.3$ Hz. Therefore, **a lumped mass and simple stiffness of a SDOFO are accurate enough for the calculation of the frequencies**. The calculated values are different from the frequencies obtained based on the measurements. A possible reason for the difference is the effect of foundation contribution according to Equation (5.2) and Fig. 4.16. The frequencies of horizontal and rocking vibration of the pile group are obtained for a stack of equivalent cylinders with the radii of 21.0 m in the longitudinal and 13.5 m in the transversal direction according to Equation (5.6). The cylinders are 12.2 m deep in a 30 m thick soil layer with transversal wave velocity of 300 m/s. The results from CONAN software are shown in Figs. 5.11 and 5.12.

From the first crest in Fig. 5.11 at frequencies greater then zero, it follows that the fundamental period of vibration of the piles in the horizontal direction is 3.2 Hz. From the first crest in Fig. 5.12 at frequencies greater than zero, it follows that the fundamental frequency of vibration of piles in rocking is 5.8 Hz in the longitudinal and 6.2 Hz in the transversal direction. From Equation (5.2), the combined frequency of vibration of the building and its foundation in the longitudinal direction is 1.7 Hz and in the transversal direction 2.4 Hz. The difference between the calculated frequencies based on the simplified models and the frequencies obtained from the measurements may be caused by the structural damage during the earthquake and its aftershocks. If cracked concrete column forms a hinge at its top or bottom then the stiffness of the columns would decrease to $3\,E\,I_x\,l^{-3}$ instead of $12\,E\,I_x\,l^{-3}$ (e.g. Jenkins, 1989) and the frequency of vibration would decrease to $(0.25)^{1/2} = 0.5$ times the frequency of vibration of the column with prevented rotations of its ends. Therefore, the frequency of vibration of a damaged structure could

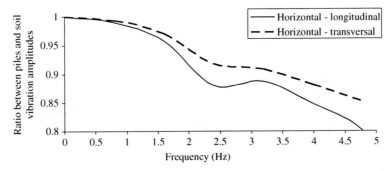

Fig. 5.11 Ratios between piles and ground vibration amplitudes in the horizontal direction in the case study in Section 5.3.4

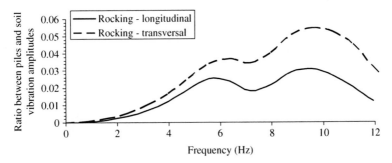

Fig. 5.12 Ratios between piles and ground vibration amplitudes during rocking in the case study in Section 5.3.4

be used for assessment of its degree of damage. For the structure with formed hinges at the column tops or bottoms, the structural frequencies would be in the longitudinal direction $0.5*2.2 = 1.1$ Hz and in the transversal direction $0.5*4.4 = 2.2$ Hz. The combined frequency of vibration of the building and its foundation would be 0.95 Hz in the longitudinal and 1.7 Hz in the transversal direction. These values are very close to the frequencies inferred from the measurements of ambient vibrations of 1.0 and 1.4 Hz respectively and therefore it would be concluded, based on calculated frequencies of its free vibrations, that the structure has suffered severe damage as it actually did.

5.4 Summary

Foundations and structures can attenuate or amplify incoming ground motion in two ways:

- 'Kinematic' interaction arises due to foundation and structure greater stiffness in comparison with the stiffness of adjacent ground. Stiff foundations and structures tend to average ground motion and in this process undergo additional stresses and strains. A simplified analysis of ground motion averaging is given in Section 5.2. Simplified analyses of additional stresses within shallow foundations are given in Section 1.4.2.1.

- Inertia interaction is caused by foundation and structure masses. Foundations and structures may oscillate with different frequencies from adjacent ground. Waves reflected from foundation will superpose with incoming waves, as indicated in Fig. 2.7, and change them. When the period of ground vibration coincides with the period of vibration of foundation and structure then (near) resonance occurs, which causes an increase in foundation and structure vibration and can cause their damage. Therefore, it is important to avoid (near) resonance between vibration of adjacent ground and foundations with structures. Simplified analysis of the amplitude amplification of vibration of a foundation and structure due to (near) resonance is given in Section 1.4.1.1.

Chapter 6
Ground Investigation for Vibration Prediction

6.1 Introduction

Data on ground profile, ground water level and ground classification properties should always be available even if attenuation relationships of ground vibration are used from literature in order to be able to assess the relevance of use of empirical data. For simplified analyses, ground and vibration source properties (energy released for body waves and maximum force amplitude for near surface waves) need to be known. Ground properties involve:

- Unit density for body waves.
- Ground damping for body waves. Ground damping is dependent both on ground type and strain (Section 2.4.7). Ground strain is dependent both on ground particle and wave propagation velocity (Equations 2.2 and 2.3) so that a recursive relationship exist between particle velocity and ground strain.
- Poisson's ratio for near surface waves.
- Stiffness modulus (or wave propagation velocity) for near surface waves.

For numerical analyses, more detailed ground properties obtainable from laboratory testing of ground samples may be required.

Field non-intrusive (geophysical) methods need to be supplemented by field intrusive methods (boreholes) in order to achieve greater accuracy and uniqueness in the interpretation of the results. Also, field intrusive methods may require their supplement by laboratory testing in order to be able to determine ground properties in the conditions different from the existing ground conditions, for example at increased shear strains and stresses. Laboratory testing of a limited number of samples represent only a tiny part of the whole mass and a representative number of samples should be obtained and tested. Sampling and testing of ground should achieve:

- Minimum sample disturbance and preservation of in situ stresses, moisture content and ambient temperature as much as possible.

M. Srbulov, *Ground Vibration Engineering*, Geotechnical, Geological, and Earthquake Engineering 12, DOI 10.1007/978-90-481-9082-9_6, © Springer Science+Business Media B.V. 2010

- Uniform stress and strain distribution within specimen during testing and ability to control or measure boundary stresses and strain.
- Stress path similar to the expected stress path in the field.
- Consistency and repeatability of the results obtained from the same test conditions.

One method of soil investigation and testing is not sufficient to provide data on all necessary ground properties and ground conditions.

The objective of this chapter is to describe basics of and comment on common/standardized methods for ground investigation and testing that may be required for prediction and analyses of ground vibrations. Other methods that can be used are described elsewhere (e.g. Kramer, 1996; Dean, 2009).

6.2 Field Non-intrusive Methods

AASHTO (2009), for example, summarizes advantage and disadvantages of geophysical surveying. The advantages are summarized as follows:

- Non-intrusive methods are beneficial when conventional drilling, testing and sampling are difficult, for example in deposits of gravel and talus, or where potentially contaminated subsurface soil may occur.
- Geophysical testing covers a relatively large area and can be used to optimize the locations and types of in-situ testing and sampling. Geophysical methods are particularly well suited to projects with large longitudinal extent compared to lateral extent, for example paths between vibration sources and the recipients.
- Geophysics assesses the properties of ground at very small strain, of the order of 10^{-5}, thus providing information on ground elastic properties.
- Geophysical methods are relatively inexpensive when considering cost relative to the large areas over which information can be obtained.

Some of the disadvantages of geophysical methods include:

- Most methods work best for situations in which there is a large difference in stiffness or conductivity between adjacent ground layers.
- It is difficult to develop good stratigraphic profiling if the general stratigraphy consists of hard material over soft material or resistive material over conductive material.
- Results are generally interpreted qualitatively and, therefore, only experienced geophysicist that is familiar with the particular testing method can obtain useful results.
- Specialized equipment is required (compared to more conventional subsurface exploration tools).

- Since evaluation is performed at very low strain, information regarding ultimate strength for evaluation of strength limit states is only obtained by correlation, which may be difficult if ground classification properties are only inferred.

6.2.1 Seismic Refraction

The method is standardized (e.g. ASTM D5777). Seismic refraction is usually used to define depths (typically to 30 m but possibly to ~300 m) of subsurface layers (usually maximum four) and ground water level as well as to determine the velocity of longitudinal wave propagation through subsurface layers. Longitudinal waves are used because they travel fastest through ground and are the first to arrive at a receiver but determination of velocity of transversal waves is possible by seismic refraction.

The method is based on measurement of travel time of longitudinal wave propagation from a source (hammer blow, weight drop, and explosive charge – more expensive) to a receiver (geophone) on the ground surface. When a longitudinal wave travelling from the source reaches an interface between two materials of different wave propagation velocities, the wave is refracted according to Snell's law (Section 2.4.2). When the angle of incidence equals the critical angle at the interface, the refracted wave moves along the interface between two materials, this interface is referred to as a refractor. From plotted times of arrival of a longitudinal wave at different distances of geophones shown in Fig. 6.1b, it is possible to calculate both the velocity of wave propagation from the slope(s) of the polygon and the depths of the refractor(s), from intercept times and crossover distances.

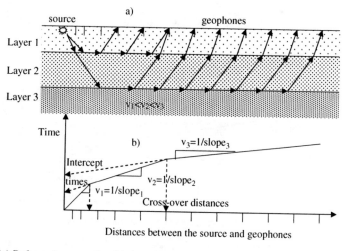

Fig. 6.1 (a) Refracted wave paths, (b) time-distance plot for three horizontal layers

The calculation formulas are based on the following assumptions:

- The boundaries between layers are planes that are either horizontal or dipping at a constant angle.
- There is no ground surface relief.
- Each layer is homogeneous and isotropic (with the same properties in any direction).
- The wave velocity increases with depth.
- Intermediate layers must have sufficient velocity contrast, thickness and lateral extent to be detected.

Ground layers can be inclined and therefore reverse measurements are used. The velocity obtained for the refractor from either of these two measurements alone is the apparent velocity of the refractor. Both, forward and reverse measurements are necessary to calculate the true wave velocity and the dip of layers unless other data indicate a horizontal layering. The wave paths and time distance plot for two layers with variable thicknesses are shown in Fig. 6.2.

The error of calculation of layer depth increases as the angle of dip of the layer increases. If a layer has wave velocity lower than the velocity of the layer above it (velocity reversal exists) then the lower wave velocity layer cannot be detected. The computed depths of deeper layers are greater than the actual depths. For irregular layer interfaces other methods exist to calculate layer profiles such as the common reciprocal method. For lateral variation in wave velocity within a layer, thin intermediate velocity layers and velocity inversion, the generalized reciprocal method is used (e.g. ASTM D5777). The seismic refraction method is sensitive to ground vibrations from various sources such as:

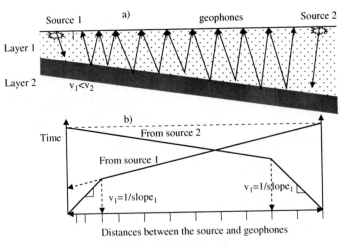

Fig. 6.2 (a) Refracted wave paths, (b) time-distance plot for a layer with dipping lower boundary

- Ambient sources of noise involve ground vibration due to wind, water movements (e.g. wave break on a beach, fluid movement in pipelines), natural micro seismicity or rain drops on geophones.
- Geologic sources of noise include lateral and vertical variations in wave velocity (e.g. presence of large boulders).
- Cultural sources of noise include vibration due to movement of the personnel conducting measurements, vehicles, construction activity, etc.

6.2.2 Seismic Reflection

The method is standardized (e.g. ASTM D7128 – 05). It is most frequently used for determination of wave propagation velocity and thickness of subsurface layers on a large scale for very deep stratigraphy and rarely for shallow soil layers. The test and its interpretation are conceptually very simple. The test is performed by using an impulse to cause usually longitudinal wave at the surface and measuring the arrival time of direct and reflected wave (Section 2.4.3) at a receiver on the surface, Fig. 6.3.

Wave velocity is calculated from direct wave and layer thickness from reflected wave assuming that wave velocity propagation is isotropic (equal in both directions). In the case of inclined layer boundary, the layer thickness and its inclination can be determined using two receivers of which one is placed at the source (e.g. Kramer, 1996), Fig. 6.4.

More than one layer can be detected and their wave velocities determined by the method. The problem arises when the times of direct and reflected wave coincide (due to wide pulse width) and also when layers have low velocities. Difficulty with resolution (Section 3.1) increases with increased source to receiver distance. Therefore, this method uses much smaller source to geophone distances in comparison with seismic refraction, which may require a source to geophone distance of up to five times the depth of investigation. McDowell et al. (2002), for example, stated that *Attempts to use higher frequency sources (giving shorter wavelengths) to improve the basic resolution of the method have been inhibited by lack of penetration*

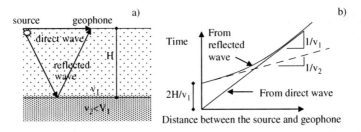

Fig. 6.3 (a) Reflected and direct wave paths, (b) time-distance plot for direct and reflected wave

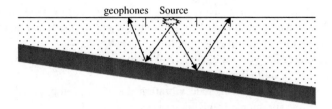

Fig. 6.4 Reflected wave paths for a layer with dipping lower boundary

of seismic pulse, caused by greater attenuation of the seismic energy in the near surface layers. Even when adequate penetration and resolution of the geological structure has been achieved, it may not be possible to observe the seismic signal if the environmental noise is excessive.

6.2.3 Spectral Analysis of Surface Waves

Development in digital data acquisition and processing equipment enables the use of spectral analysis of ground surface response to near surface propagating Rayleigh waves for determination of variation of body wave's velocity with depth based on measurements of ground surface velocity by geophones at 4–100 Hz or acceleration at 100–5000 Hz. A source of varying vibration frequency is used to obtain the shape of a dispersion curve, which is a plot of Rayleigh wave velocity versus frequency. The recordings from minimum two receivers are transformed to the frequency domain using the fast Fourier transform and the phase difference and corresponding wave travel time are computed for each frequency. This eliminates the problem of detecting wave arrivals and measuring arrival times. From known distances between the receivers, Rayleigh wave phase velocity and wavelength are calculated as functions of frequency.

Identification of the thickness and transversal wave velocity of subsurface layers involves the iterative matching of a theoretical and the experimental dispersion curves (e.g. Kramer, 1996). The theoretical dispersion curve is available for horizontal layers and therefore the disadvantages of the method are:

- It is limited to sites with near horizontal layering.
- Specialized equipment and experienced operators are required to perform the test.

The advantages of spectral analyses of surface waves are (e.g. Kramer, 1996):

- The test is quick to perform and it is possible to obtain the results in the field in real time using modern electronic instrumentation.
- No boreholes are required.
- Can detect low velocity layers.
- Can be used to considerable depth (>100 m).

Tallavo et al. (2009) concluded that multiple channel analysis of surface waves can be successfully used to define the surface location of decayed buried trestles. The idea to use the method came probably from its use for detecting of underground voids, as referenced in their paper.

6.2.4 Seismic Tomography

The geophysical methods described in Sections 6.2.1–6.2.3 are applicable to near horizontal ground surface and/or subsurface layers. Seismic tomography can be used to infer distribution of ground wave velocities within zones of any shape by using arrays of vibration sources and receivers as sketched in Fig. 6.5.

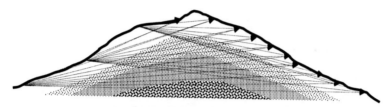

Fig. 6.5 Locations of vibration sources and receivers (geophones) for obtaining the image of interior zones within an anticline ridge

The method requires an extensive data gathering and computer processing and therefore is expensive to use for routine tasks.

6.2.5 Ground Penetrating Radar

Besides geophysical methods mentioned in Sections 6.2.1–6.2.4, which are based on mechanical ground wave propagations (seismic methods), other methods exist (e.g. ASTM D6429-99). One of them, the surface ground penetrating radar is standardized (e.g. ASTM D6432-99).

Ground penetrating radar uses reflection of high frequency electromagnetic waves (from 10 to 3000 MHz) for detection of interfaces within ground with different electromagnetic properties. Not only that soil and rock have different electromagnetic properties but also buried materials such as unexploded ordnance. The penetration is usually less than 10 m in most soil and rock. Penetration in clay and in conductive pore fluids may be less than 1 m. However, the radar provides the highest lateral and vertical resolution (at high frequencies but low penetration) of any geophysical method conducted from ground surface. It can be used to locate small targets such as steel rebar in reinforced concrete. McDowell et al. (2002) consider that for detecting voids in rock *Radar is potentially the most useful technique as it provides the highest resolution and, in good conditions, can penetrate to considerable depth.*

6.2.6 Field Compaction

Mooney and Rinehart (2007) reported on a field investigation that was carried out with and instrumented vibratory roller compactor to explore relationship between vibration characteristics and soil stiffness. *Using lumped parameter vibration theory, soil stiffness was extracted from the roller data (drum and frame acceleration and drum phase lag). Both drum acceleration and drum phase lag were found to be very sensitive to changes in underlying soil stiffness. The drum-soil natural frequency of the coupled roller-soil system varied considerably and increased with compaction soil stiffening. Roller determined soil stiffness was found to be a function of the eccentric force, and heterogeneity in moisture, lift thickness, and underlying stiffness has a considerable effect on roller vibration behaviour.*

6.3 Field Intrusive Methods

Field non-intrusive methods require confirmation of the ground layering inferred by the use of the methods from borehole loggings. The same boreholes can be used also to obtain measurements of ground wave propagation.

6.3.1 Seismic Down-Hole

The method is standardised (e.g. ASTM D7400-08). It can be performed in a single borehole when a vibration source is located on the ground surface and a single receiver moved to different depths or a number of receivers are fixed at different depths in the borehole. From a plot of measured travel time of longitudinal or transversal waves versus depth, ground wave propagation velocity at a depth is obtained from the slope of the plot (e.g. Kramer, 1996), Fig. 6.6.

In a version called up-hole test, a movable energy source is located in borehole with a single receiver on the ground surface close to the borehole. However, transversal waves can be generated much more easily when the source is on the ground surface and, therefore, down-hole test is more frequently used. The down-hole test allows detection of layers that can be undetected in seismic refraction survey. Also, it provides ground wave velocity in the vertical direction, which may be different from velocities in the horizontal and slant direction in anisotropic soil. Kramer (1996), for example, mentions potential difficulties with down/up-hole test and their interpretation as:

- Soil disturbance by drilling equipment.
- The use of casing and borehole fluid for borehole stabilisation.
- Insufficient or excessively large impulse source.
- Background noise effects.

Fig. 6.6 (a) Cross section of
a seismic down-hole test, (b)
travel-time versus depth
graph

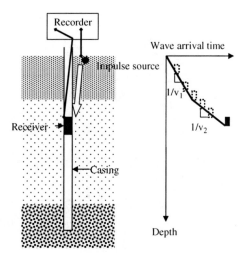

- Groundwater level effect.
- The effects of material and radiation damping on waveforms can make identification of transversal wave arrivals difficult at depth greater than 30–60 m.

6.3.2   Seismic Cross-Hole

The method is standardized (e.g. ASTM D4428/D4428M-00). The method is used for determination of velocity of longitudinal and transversal ground waves propagating horizontally through primarily soil layers, with absence of rock. Minimum two but preferably three boreholes are used in line spaced at 3 m apart, in order to minimize chance that the arrival of refracted rather than direct waves is recorded. When a higher velocity horizontal layer exists near the level of the source and receivers then the recorded wave amplitude may be the result of a refracted wave propagated through the higher velocity layer. ASTM D4428/D4428M–00, for example, provide example calculations for this case. PVC pipe or aluminium casing is grouted in boreholes and three-directional receivers (geophones or accelerometers) must be fixed firmly to them. Impulsive source used can be explosive charge, hammer or air gun. A sketch of cross section through boreholes is shown in Fig. 6.7.

The cross-hole test allows detection of layers that can be undetected in seismic refraction survey. Also, it provides ground wave velocity in the horizontal direction, which may be different from velocities in the vertical and slant direction in anisotropic soil. The test results (i.e. measured times of incoming waves at the receivers) may be affected by borehole deviation and therefore a survey must be conducted to establish true horizontal distances between boreholes, particularly at depths greater than 15–20 m. Wave amplitude attenuation measurements from cross-hole tests performed in at least three boreholes has been used to obtain

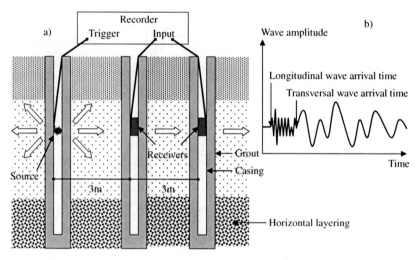

Fig. 6.7 (a) Cross section of the test setup, (b) recorded wave amplitudes from a reversible impulse source

material damping ratios at small strain (e.g. Kramer, 1996). Boreholes are used not only for measurements of ground wave propagation velocity and detecting of ground layering but also for taking samples for laboratory testing and performing of in situ tests. An in-situ test called standard penetration test (SPT) (e.g. ASTM D1586 – 08a; Eurocode 7 – Part 2, 2007) is performed in boreholes world wide and routinely. Japan Road Association (2003), for example, provides expressions for assessment of soil transversal velocity based on the average blow count N_i of SPT in a soil layer:

$$\begin{aligned} &\textit{For cohesive soil layer}: v_t = 100 \cdot N_i^{1/3}, \ 1 \leq N_i \leq 25 \\ &\textit{For sandy soil layer}: v_t = 80 \cdot N_i^{1/3}, \ 1 \leq N_i \leq 50 \end{aligned} \qquad (6.1)$$

6.3.3 Seismic Cone

It is an addition to the electric piezo-cone (e.g. ASTM D5778-95; Eurocode 7 – Part 2, 2007) and is described by Lunne et al. (2001). Figure 6.8 shows a sketch of the cone.

As standard cone (35.7 mm diameter i.e. 10 cm^2 cross sectional area and 60° apex angle) can be used for soil classification and identification of ground layering, seismic cone can be used instead of seismic down-hole test or as an addition to it because the use of piezo-cone is fast (penetration speed 2 ± 0.5 cm/s) and may be cheaper. Its use is limited by the availability at a site and depth of penetration that can be achieved with pushing in force of up to 200 kN, depending of vehicle on which the cone is mounted.

Fig. 6.8 Sketch of an electric
piezo seismic cone

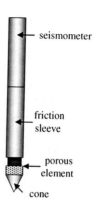

Miniature velocity seismometer is usually placed above a piezo-cone unit. Other necessary equipment includes a memory oscilloscope and an impulse source, which is capable of generating high frequency waves, located on the ground surface (at less than 1 m from the cone push-in rod). Except this downward wave propagation setup, a version was used involving two seismic cones in parallel, one as a source and the other as a receiver, similar to seismic cross-hole test described in Section 6.3.2. In such a setup, the verticality of the cone rods is important. Both longitudinal and transversal waves can be generated on the ground surface. The measurements are usually performed at 1 m interval during a brief break in a continuous penetration. The use of two receivers 0.5 or 1 m apart can improve the quality of the recorded data by eliminating problems related to triggering times.

Pushing in of cone penetrometer into ground causes inevitably formation of a zone of remoulded ground and large strain deformation around the cone rod. The extent of this zone is relatively small in comparison with the whole source to receiver distance passing through undisturbed ground.

6.4 Laboratory Testing

These tests are used when ground vibration induced stress and strain states are expected to be different from the existing states within ground. For example, pile driving, soil compaction and blasting are expected to induce large strains in the vicinity of the sources in comparison with small strain (less than 10^{-6}) induced by propagation of ground waves during field testing. Large strains are associated with ground failure and, therefore, laboratory tests are really necessary only for investigation of ultimate equilibrium condition and not for considerations of serviceability criteria. The order of description of various laboratory tests corresponds to the frequency of their use.

6.4.1 Bender Elements

The elements have been in use for the last 40 years since their description by Shirley (1978). The following description is based on Dyvik and Madshus (1985). The piezo-ceramic bender element is an electro-mechanical transducer, which is capable of converting mechanical energy either to of from electrical energy. The element consists of two thin piezo-ceramic plates (about 0.5 mm thick, 10 mm wide and 12 mm long depending on the place of application), which are rigidly bonded together with conducting surfaces between them and on the outsides. When electrical voltage is applied to the element, one plate elongates and the other shortens resulting in their bending and emission of outgoing transversal ground wave. When incoming ground wave bend the element, this results in generation of electrical signal, which can be measured. Depending on the connection between the conducting surfaces, a series and a parallel connection are used as a receiver (mechanical to electrical energy) and transmitter (electrical to mechanical energy) converters respectively based on the connection effectiveness. The elements operate at small strain only and are useful not only for an independent measurement of transversal wave velocity at small strain but also for checking of the results of other tests used for large strain but at the initial small strain stage. To prevent current shortcut due to presence of moisture, the elements are cased in a waterproofing material. Cracks in the waterproofing can result in a significant dropping of impedance of the transmitter, which should be checked regularly. The movement of the element should be free and therefore water resistant sealants should be used around it as sketched, for example, in Fig. 6.9.

Transversal wave velocity is obtained as the ratio between tip to tip distance of a top and bottom bender element and the elapsed time for the transversal wave to pass

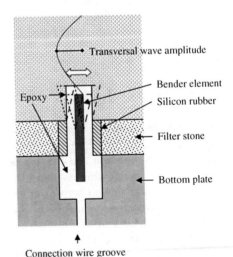

Fig. 6.9 Cross section through a bottom mounted bender element

through soil specimen. A number of research studies have been conduced over the years of use of bender elements. For example, Arroyo et al. (2006) investigated *the heavily distorted transmission usually observed when the test interpretation is based on the assumption of plane wave propagation between transducers* and consider that *a main culprit for signal distortions is sample-size effects due to lateral boundary reflections.* Lee and Santamarina (2005) explored various aspects of bender element installations and find that:

- Electromagnetic coupling effects are critical in soils with high electrical conductivity and can be minimized by shielding and grounding or by using parallel-type bender elements.
- The in-plane transversal wave directivity is quasi circular.
- The resonant frequency of bender element installations depends on the geometry of the bender element, the anchor efficiency, and soil stiffness.
- The cross correlation of subsequent reflections is a self-healing measurement procedure, which resolves uncertainties in both travel time and travel distance.
- Near field effects can be effectively taken into consideration by matching the measured signal with the analytical solution, directly rendering transversal wave velocity.

An ASTM committee is working on standardization of bender element test.

6.4.2 Cyclic Simple Shear

The test and its typical results are described by Finn (1985), for example. Cyclic simple shear apparatus is used frequently for testing of potential to liquefaction of saturated sandy soil because it approximates closely usual assumptions introduced by Seed and Idriss (1967) that:

- Seismic excitation is due primarily to transversal waves propagating vertically.
- Level ground conditions may be approximated by horizontal layers with uniform properties.

The apparatus is also used for the investigation of degradation of the strength and stiffness of saturated clay in cyclic condition. Cyclic simple shear test consists of applying a cyclic force or displacement to a top or bottom surface of thin ground specimen, with width to height ratio in excess of 5 preferably. Constant cyclic force/displacement amplitude is used when an equivalent number of significant stress cycles concept (Seed et al., 1975) is considered. Time histories of irregular cyclic force/displacement amplitudes and in two directions may be applied as well (e.g. Ishihara and Nagase, 1985). The first simple shear test apparatus was developed by Roscoe (1953) for static testing of soil. It was adapted to cyclic loading conditions first by Peacock and Seed (1968). The apparatus development follows

Fig. 6.10 Sketch of cross section through Roscoe simple shear apparatus

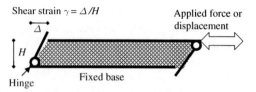

recognition of the deficiencies of direct shear test device, in which a ground specimen is enclosed within rigid metal box that is split in the middle and is capable of shearing the specimen along a distinct shear zone.

The Roscoe apparatus uses a specimen of rectangular cross-section and generates simple shear (change of shape by distortion) under plane strain conditions, as shown in Fig. 6.10 in the case when constant volume is maintained so that no displacement occurs in direction perpendicular to the specimen thickness. Therefore, the sloping lengths of the sides of the test specimen must increase as shear progresses.

The slipping required to accommodate the change in side plate length violates an ideal boundary condition for simple shear at the ends as complementary shear stresses cannot be developed. The shear stresses that occur during slip are indeterminate so to ensure controlled conditions during a test these frictional shear stresses should be reduced as much as possible, ideally to zero by lubricating the side plates. Other simple shear devices exist; the Norwegian Geotechnical Institute uses a wire-reinforced membrane around a cylindrical specimen while the Swedish Geotechnical Institute uses a series of staked rings (e.g. Kramer, 1996). When shared in these devices, the specimen does not maintain plane strain conditions (Finn, 1985). In Roscoe apparatus, essentially uniform simple shear conditions can be induced in most parts of a specimen particularly at small strain. Around the boundaries and near the ends, narrow zones of smaller than average shear strain exist.

The test results are plotted as a graph shown in Fig. 2.9. From the shear modulus G_{secant} obtained from the test results, it is possible to back calculate transversal wave velocity at a particular strain level as a square root of the ratio between the shear modulus at that strain level and soil unit density. A damping ratio is calculated from Equation (2.32). Ishihara and Nagase (1985) summarized the effects of multidirectional and irregular loading on the tests results as follows:

- The maximum shear stress to the initial confining stress ratio at 3 to 5% shear strain in the multidirectional loading test is about 1.5 times the cyclic stress ratio necessary to cause the same range of shear strain in unidirectional uniform loading test for loose sand (with density of 45% of the maximum value) and for medium dense sand (with density of 75% of the maximum value) and 0.9 times the stress ratio in unidirectional uniform loading test for dense sand (with density of 95% of the maximum value). In other wards, loose and medium dense sand

are more difficult and dense sand more easily disturbed by multidirectional than unidirectional uniform loading shaking.
- While the effects of multi-directivity of loading are not influenced significantly by sand density, the effects of load irregularity are manifested to a varying degree depending on sand density.

6.4.3 Cyclic Triaxial Test

The test is standardized (e.g. ASTM D3999 – 91). The cyclic triaxial test consists of applying either a cyclic axial stress of fixed magnitude (load control) or cyclic axial deformation (displacement control) on a cylindrical soil specimen enclosed in a triaxial pressure cell. The specimen diameter could vary from 38 to 300 mm and the height to diameter ratio should not be less than 2 in order to minimize non-uniform stress conditions within the test specimen, which are imposed by the specimen end platens, Fig. 6.11.

Although the test is used for strain level up to 0.5%, loose sand/ soft clay tend to experience bulging in the middle of the specimen and dense sand/ stiff clay tend to develop a distinct inclined shear zone, which cause a highly non-uniform distribution of strain within the specimen. Local measurements of strain are performed sometimes on the specimen side within the cell particularly when the specimen response at small strain is investigated. Such measurements are not standardized and are performed by specialized soil testing laboratories (e.g. Jardine et al., 1985; Burland, 1989; Tatsuoka et al., 1997). Such specialized tests are more expensive but also the cost of cyclic triaxial test is about ten times greater than the cost of static triaxial test.

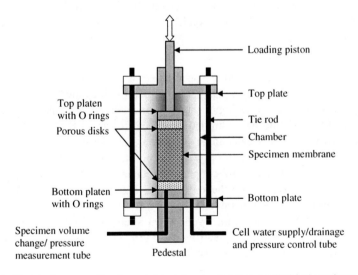

Fig. 6.11   Sketch of cross section through triaxial pressure cell and cylindrical specimen

The cylindrical specimens have been obtained from nearly vertical boreholes and therefore the test simulates best the effects of propagation of nearly vertical longitudinal waves. The test results are plotted as a graph of axial stress versus axial strain similar to the graph of shear stress versus shear strain shown in Fig. 2.10. From the secant axial modulus obtained from the ratio between axial stress and strain, it is possible to back calculate longitudinal wave velocity at a particular strain level as a square root of the ratio between the axial modulus at that strain level and soil unit density. A damping ratio is calculated from Equation (2.32). Other features of the test are:

- A 90° change in the direction of the major principal stress occurs during the two halves of the loading cycle. If cyclic deviator (axial less radial) stress is smaller than the deviator stress during consolidation, no stress reversal occurs.
- The maximum cyclic axial stress that can be applied to the specimen is equal to the effective confining pressure (cell pressure less the excess pore water pressure within the specimen).
- Different methods of preparing specimen from disturbed soil sample may result in significantly different cyclic behaviour.
- Membrane compliance effect cannot be simply accounted for in the test procedure or in interpretation of test results. Changes in pore water pressure can cause changes in membrane penetration in specimens of coarse cohesionless soils. These changes can significantly influence the test results.
- Degree of specimen saturation can significantly affect the axial stiffness and consequently the longitudinal wave velocity inferred from the axial stiffness particularly at near saturation state. For saturated soil the longitudinal wave velocity correspond to about 1500 m/s, i.e. longitudinal wave velocity propagation through water, and for degree of saturation of less than about 99%, the longitudinal wave velocity corresponds to the wave velocity through dry soil (e.g. Gazetas, 1991).

The results of tests performed on specimens in axi-symmetric strain condition in cyclic triaxial apparatus and in plane strain condition in cyclic simple shear apparatus may be different. For testing of liquefaction potential, for example, the results from triaxial tests should be multiplied by a factor of about 0.7 (Seed, 1979). For static condition however, greater angle of friction of sand is measured in plane strain condition than in triaxial condition (e.g. Cornforth, 1964). Townsend (1978), for example, summarized factors affecting cyclic triaxial strength of cohesionless soil. Similar effects are expected on the soil stiffness and wave velocity. The effects of influential factors are listed in Table 6.1.

The effects of factors affecting cyclic shear strength of normally consolidated clay are summarized by McClelland Engineers (1977). Similar effects are expected on the soil stiffness and wave velocity, Table 6.2.

Table 6.1 Effects of most influential factors on the cyclic triaxial strength of cohesionless soil

Factor	Effect
Specimen preparation method	Weakest specimens formed by pluviation through air, while strongest ones formed by vibrating in moist condition. Difference in stress ratio at failure can be 110%.
Reconstituted versus intact	Intact specimens stronger than reconstituted. Strength decrease range from 0 to 100%.
Confining stress	Cyclic strength is directly proportional to confining stress within small range of pressure. Cyclic stress ratio decreases with increasing confining pressure.
Loading waveform	Strength increases from rectangular wave shape, via triangular to sine. Sine wave causes approximately 30% greater strength than rectangular. Irregular wave form can be replaced by equivalent harmonic wave.
Frequency	Slower loading frequencies have slightly higher strength. For a range from 1 to 60 cycles per minute, the effect is 10%. Water presence may affect results at 5 cycles per second.
Specimen size	300 mm diameter specimen exhibit approximately 10% weaker strength than 70 mm diameter specimen.
Relative density	Exponential shear strength increase with linear increase in density.
Particle size and degradation	Sand with average diameter D_{50} of approximately 0.1 mm has least resistance to cyclic loading. As D_{50} increases from 0.1 to 30 mm, shear strength increases 60%. As D_{50} decreases from 0.1 mm to silt and clay sizes, a rapid increase in strength is observed. Well graded soil weaker than uniformly graded soil.
Pre-straining	Previous cyclic load greatly increases shear strength during current cyclic load.
Over consolidation	Over consolidation increased shear strength depending on amount of fines (particles less than about 5 μm).
Anisotropy	Shear strength is increased by increased anisotropy. Method of data presentation influences the effect; isotropic consolidation may not always provide conservative results.

Table 6.2 Factors affecting cyclic strength of normally consolidated clay

Factor	Change in factor	Change in undrained shear strength
Cyclic stress	Increase	Decrease approximately linearly with the logarithm of number of cycles
Number of stress cycles	Increase	Decrease
Initial shear stress	Increase	Decrease
Direction of principal stress	90° rotation	Decrease
Shape of cyclic stress	From square to sine	Decrease
Frequency of cyclic stress	From 2 to 1 cycles per second	Decrease
Stiffness of soil	Increase	Increase
Stress state	From triaxial to simple shear	Negligible

Poisson's ratio can be obtained from static triaxial tests. The use of Mohr circle for a comparison of the results of triaxial and simple shear test is demonstrated by Dean (2009), for example.

6.4.4 Resonant Column

The test is standardized (e.g. ASTM D4015-92). The test consists of vibrating a cylindrical specimen in torsion to determine shear modulus and damping ratio or in axial direction to determine axial modulus and damping ratio until resonance between the excitation and the system vibration is achieved. The specimen diameter could vary from 33 mm onwards and the height to diameter ratio should not be less than 2 or more than 7 except when applied axial stress is greater than confining stress in which case the ratio shall not be greater than 3. Nevertheless, non-uniform stress condition will exist in torsion with the average shear strain for each cross section to occur at a radius equal to 80% the radius of the specimen. Different types of apparatus exist. Skoglund et al. (1976), for example, compared the results obtained by six different investigators and found that the differences in shear and axial modules ranged from minus 19% to plus 32% of the average value. No systematic or consistent differences could be associated with different types of apparatus used. A variant of test schematic with fixed top end and movable bottom end is shown in Fig. 6.12 without a triaxial chamber and loading piston.

Test can be performed at different strain but it is considered to be non-destructive when strain amplitude of vibration is less than 10^{-4} (radians). Details of calculation of modulus and damping ratio depend on the apparatus used. Clayton et al. (2009), for example, investigated the effects of apparatus stiffness and mass, and specimen fixity, on the results obtained from Stokoe apparatus with free top end. They showed that the stiffness of the drive head and the mass of the apparatus base can both have a significant effect on the results obtained when stiff specimens are tested in addition to issues concerning connectivity between the specimen, the apparatus base and the top cap.

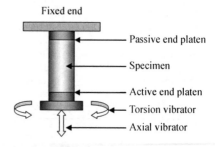

Fig. 6.12 Schematic of a resonant column test with fixed top end without triaxial chamber and loading piston

6.5 Summary

Data on ground profile, ground water level and ground classification properties should always be available even if attenuation relationships of ground vibration are used from literature in order to be able to assess the relevance of use of empirical data. For better understanding and prediction of ground vibration, ground profile and properties can be inferred using a number of methods:

- Field non-intrusive methods are relatively fast and inexpensive but the results are unreliable when stiffer layers overly softer layers and need knowledge of ground profile. Geophysical (seismic) methods are applied at small strain, while large strain exists in vicinity of pile driving, soil compaction or blasting.
- Field intrusive methods are more expensive and time consuming then non-intrusive methods but may be applied when stiffer layers overly softer layers. The electric piezo and seismic cone penetration test is capable of replacing drilling of borehole and performing geophysical testing in them but is limited by the depth of cone penetration.
- Laboratory testing is applicable at large strain and higher ambient stresses than strain and stress existing in the field when using geophysical (seismic) methods.

Ground investigations should be used to establish relationships between ground damping versus strain, ground wave velocity or stiffness versus strain, Poisson's ratio and unit density at least for the ground layers close to ground surface.

Chapter 7
Prediction of Vibration Amplitudes

7.1 Introduction

Prediction of ground vibration amplitudes and their comparison with legislative values is a frequent engineering task. Different methods for prediction of ground vibration amplitudes exist:

- Empirical methods are widely used in practice (as shown later in this chapter) and are based on available attenuation relationships of measured ground vibration amplitudes with distance from the source. Problem with their use is that the existing attenuation relationships may not be available and if they are then complementary data about ground condition and vibrating source to which they apply may not be specified so that it is difficult to assess their relevance to a problem at hand.
- Simplified analyses can always be used but they require knowledge of properties of vibration source and of ground conditions, which may not be available in part or in total.
- Numerical analyses are expected to provide accurate solutions of the problem but they require the use of proprietary software (e.g. listed in http://www.ggsd.com), expertise in its use and frequently detailed ground properties. Lack of expertise and/or detailed ground properties affect the accuracy of the results of numerical methods.
- Small and full scale testing is most convincing method for assessment of future ground vibration amplitudes but requires the use of a specialist laboratory or field instruments and the expertise. Therefore, the testing is not so frequently used in practice.

A compromise between expected accuracy, cost, required expertise and time consumption can be achieved using simplified analyses. Wolf (1994) proposed that simplified models should satisfy several requirements:

- They should offer conceptual clarity and physical insight.
- They should be simple in physical description and in application, permitting an analysis with a hand calculator or a spreadsheet in many cases.

M. Srbulov, *Ground Vibration Engineering*, Geotechnical, Geological, and Earthquake Engineering 12, DOI 10.1007/978-90-481-9082-9_7,
© Springer Science+Business Media B.V. 2010

- Yet they should have sufficient scope of application (for different shapes, soil profiles, ground properties).
- They should also offer acceptable accuracy, as demonstrated by comparing the results of the simplified models with those of rigorous methods and prototypes.
- They should be adequate to explain the main physical phenomena involved, and have direct use in engineering practice for everyday design.
- They need to be useable for checking the results of more sophisticated analyses.
- Finally, there should be a potential generalization of the concept with clear links to the rigorous methods.

The objectives of this chapter is to present some available empirical relationships of peak particle velocities attenuation with distance to vibration sources, to provide simple calculations of peak particle velocities (PPV) and to compare calculated with recorded PPV from a number of case histories of construction/demolition, traffic and machinery operation.

7.2 Construction and Demolition Caused Vibration

7.2.1 Pile Driving

A brief description of the problem is provided in Section 1.3.1.1. It is usual to start considerations of expected ground vibration amplitudes at a location with reference to relationships of vibration amplitudes attenuation with distance from a vibration source based on empirical data as shown in Figs. 7.1, 7.2 and 7.3. Also, a number of empirical peak particle velocity attenuation relationships exist. For example, BS 5228-2 (2009) provides the following relationships for the resultant peak particle velocity (mm/s):

- For vibratory hammers

$$v_{res} = \frac{k_v}{x^{1.3}}, \tag{7.1}$$

where $k_v = 60$ (for 50% probability of exceedance), 136 (for 33.3% probability of exceedance) and 266 (for 5% probability of exceedance); x is the horizontal distance along the ground surface in the range from 1 to 100 m.
- For impact hammers

$$v_{res} \leq k_p \frac{\sqrt{W}}{r_s^{1.3}}, \tag{7.2}$$

where W is the nominal energy (J) of an impact hammer in the range from 1.5 to 85 kJ, r_s is the radial (slant) distance (m) between source and receiver (for the pile depth range from 1 to 27 m and the horizontal distance along the ground surface range from 1 to 111 m); $k_p = 5$ at pile refusal, otherwise in the range from 1 to 3 for loose to very stiff/dense soil.

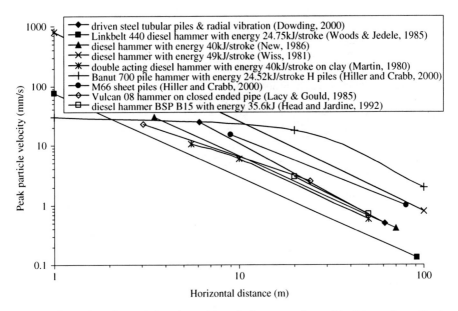

Fig. 7.1 Examples of recorded peak particle velocity attenuations with distance from diesel hammers

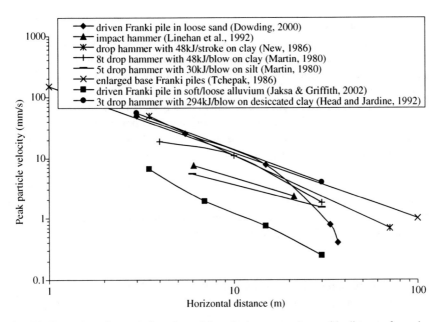

Fig. 7.2 Examples of recorded peak particle velocity attenuations with distance from drop hammers

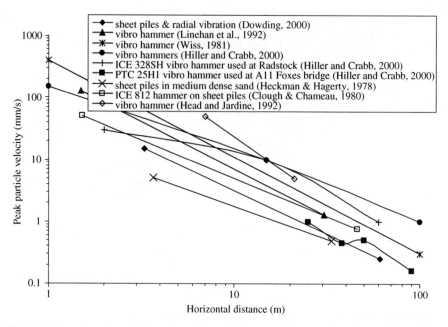

Fig. 7.3 Examples of recorded peak particle velocity attenuations with distance from vibratory hammers

A considerable scatter of peak particle velocities in Figs. 7.1, 7.2, and 7.3 is caused by the differences in energies released at the sources and ground conditions along wave propagation paths. Figures 7.1, 7.2 and 7.3 indicate that peak particle velocity is likely to exceed the peak velocities specified in various codes (Section 1.2) at vibration source to location distances of about 10 m and that more case specific simplified calculations may be required. The peak particle velocity of ground surface caused by the effect of pile shaft driving by impact and vibratory hammers can be calculated according to Equation (2.11). The peak particle velocity of ground surface caused by pile toe penetration through ground can be calculated using Equation (2.8).

Calculation of the source energy E_o in Equations (2.8) and (2.11) should be straight forward providing that the basic properties of impact hammer used and of the adjacent ground are known. A scale of machinery used for pile driving can be recognized from Fig. 7.4, for example.

7.2.1.1 Calculation of Source Energy E_o Due to Pile Driving in the Simple Analyses

The source energy E_o is product of resultant force acting along pile shaft and at pile toe and the ground displacements along pile shaft and at pile toe respectively that are

Fig. 7.4 820 mm diameter
steel tubular pile positioned in
place before driving in
Azerbaijan

caused by such forces. Calculated energy E_o cannot exceed the rated energy of pile
hammer multiplied by the hammer efficiency factor. For tubular unplugged piles,
the internal shaft friction causes the vibration of the internal soil column within the
pile and not of soil surrounding the pile.

In coarse grained soil without cohesion (i.e. with zero shear strength at zero
effective stress):

- The force along pile shaft is commonly calculated as:

$$\sigma'_{v,avr} \cdot K_s \cdot \tan \delta_\phi \cdot D_p \cdot \pi \cdot L_p, \tag{7.3}$$

where $\sigma'_{v,avr}$ is an average effective overburden pressure along pile shaft length
L_p in ground, K_s is the coefficient of lateral effective stress acting on pile shaft, δ_ϕ
is friction angle between ground and pile shaft (usually assumed equal to about
2/3 of the ground friction angle for pre-cast driven piles and equal to ϕ for cast in
place piles using compaction of concrete by a weight drop inside the retrieving
steel tube of Franki type piles), and D_p is pile diameter. For sheet pile walls,
$D_p\, 2\, L_p$ is used instead of $D_p\, \pi\, L_p$ in Equation (7.3), where D_p is the width of
the wall member being driven by a hammer.
- The force at the toe of a plugged pile is commonly calculated in non-cohesive
ground as:

$$\sigma'_v \cdot N_q \cdot \frac{D_p^2 \cdot \pi}{4}, \tag{7.4}$$

and of an unplugged pile as:

$$\sigma'_v \cdot N_q \cdot (D_p - d_p) \cdot \pi \cdot d_p, \tag{7.5}$$

where σ'_v is effective overburden pressure at the level of pile toe, N_q is ground bearing capacity factor, D_p is external pile diameter, d_p is pile wall thickness for hollow piles. For sheet pile walls, the toe force is much smaller in comparison with the side force and can be ignored.

In fine grained soil with cohesion in fully saturated and undrained condition (i.e. when there is no time for soil consolidation to take place under applied load):

- The force along pile shaft is commonly calculated as:

$$\alpha_p \cdot c_{u,\,ravr} \cdot D_p \cdot \pi \cdot L_p, \tag{7.6}$$

where α_p is ground cohesion mobilization factor along pile shaft, $c_{u,avr}$ is an average ground cohesion in undrained condition along pile shaft length L_p in ground, and D_p is pile diameter. For sheet pile walls, $D_p\, 2\, L_p$ is used instead of $D_p\, \pi\, L_p$ in Equation (7.6), where D_p is the width of the wall member being driven by a hammer.
- The force at the toe of a plugged pile is commonly calculated in cohesive ground as:

$$9 \cdot c_u \cdot \frac{D_p^2 \cdot \pi}{4}, \tag{7.7}$$

and of an unplugged pile as:

$$9 \cdot c_u \cdot (D_p - d_p) \cdot \pi \cdot d_p, \tag{7.8}$$

where c_u is cohesion of ground under/around pile tip. For sheet pile walls, the toe force is much smaller in comparison with the side force and will be ignored.

In layered ground containing layers of cohesive and non-cohesive, Equation (7.3) is applied along non-cohesive and Equation (7.6) along cohesive layers.

Cone penetrometer (Section 6.3.3) is a scaled model of a pile. Forces acting along a pile shaft and at the pile toe can be calculated by applying the scaling factor of $D/3.57$ to the force acting along the shaft of cone penetrometer and $0.25 * \pi D^2\, (10)^{-1}$ to the force acting at cone penetrometer tip, where the pile diameter D is in cm and pile squared diameter D^2 is in cm^2, the cone diameter is 3.57 cm and the cone cross sectional area is 10 cm^2.

For non-geotechnical engineers, further comments on determinations of ground parameters used in Equations (7.3) to (7.8) may be useful.

- K_s – **coefficient of effective stress acting on pile shaft** varies in the rage from 1 to 2 of K_o for large displacement driven piles or from 0.75 to 1.75 of K_o for small displacement driven piles (from Table 7.1 in Tomlison, 2001). Large displacement piles are for example sand compaction and Franki piles, which cause lateral compaction of adjacent soil due to the increase of pile diameter during pile driving. Lower values of K_s correspond to loose to medium dense and upper values of K_s to medium dense to dense soil. K_o is the coefficient of soil lateral effective stress at rest and is typically approximated as $1-\sin\phi$, following Jaky (1944), for normally consolidated soils, where ϕ' is soil friction angle. In over consolidated soil with over consolidation ratio OCR, $K_o = (1-\sin\phi') \times OCR^{\sin\phi'}$. Over consolidation ratio is between previous effective overburden pressure and existing overburden pressure. Apparent OCR can be caused by soil cementation, desiccation and by secondary consolidation.
- **Soil friction angle ϕ** and soil-pile shaft angle δ_ϕ can be determined from laboratory tests of soil samples. Peck et al. (1974) proposed a correlation between angle ϕ and the standard penetration test (SPT) blow count N_{SPT}. For the range of N_{SPT} from 10 to 40, for medium dense to dense sand, their results can be expressed as:

$$\phi = 30 + \frac{10}{35} \cdot (N_{SPT} - 10) \tag{7.9}$$

Hungr and Morgenstern (1984) found slight effect of rate of shear dependence of shear strength of sand. However, loose and medium dense sand ($0 < N_{SPT} < 30$) tends to develop positive excess pore water pressure with an increase of number of cycles. Seed et al. (1985) defined boundaries of cyclic stress ratio $\tau(\sigma_v')^{-1}$ ($= \tan\phi$ in cyclic condition) between liquefied and non-liquefied sandy soil during magnitude 7.5 earthquakes, which cause strong ground motion similar to the motion induced in vicinity of pile driving. Friction angles ϕ in cyclic conditions from these boundaries and Equation (7.9) that is applicable to the case of no excess pore water pressure development (linear parts of the graphs) are shown in Fig. 7.5. The fines in Fig. 7.5 mean content by weight of soil particles with diameters less than 0.06–0.075 mm.

Different proposals exist for correction of $\tan\phi$ for earthquake magnitudes M different from 7.5. For example, Eurocode 8 – part 5 recommends a correction proposed by Ambraseys (1988), which is closely approximated by the polynomial $0.154M^2 - 2.94M + 14.34$. Others, like Youd and Idriss (2001), for example, suggest the relationship for correction of $\tan\phi$ for earthquake magnitude $M = 7.5$ shown in Fig. 7.5 by factor $10^{2.24}M^{-2.56}$.

A measured SPT blow count N_{SPT} can be normalized to an overburden pressure of 100 kPa according to Liao and Whitman (1986), and can be corrected to an energy ratio of 60% (the average ratio of the actual energy E_m delivered by hammer to the theoretical free-fall energy E_{ff})

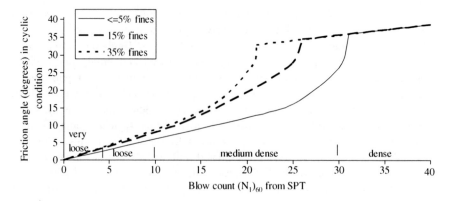

Fig. 7.5   Sandy soil friction angles in cyclic condition

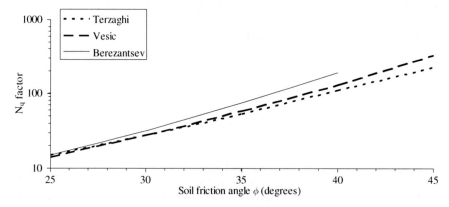

Fig. 7.6   Bearing capacity factor N_q versus soil friction angle ϕ in degrees

$$(N_1)_{60} = N_{SPT} \cdot \sqrt{\frac{100}{\sigma'_v}} \cdot \frac{E_m}{0.6 \cdot E_{ff}}, \sigma'_v \ is \ in \ \text{kPa}$$

$$0.5 < \sqrt{\frac{100}{\sigma'_v}} < 2, \quad (\text{Eurocode } 8-5) \tag{7.10}$$

$$N'_{SPT} = 0.75 \cdot N_{SPT} \ at \ \text{depths} \leq 3m, (\text{Eurocode } 8-5),$$

where σ'_v is the effective overburden pressure at the depth where N_{SPT} is recorded from SPT's blow count. Other corrections to N_{SPT} are applied, such as for the borehole diameter, rod length and sampler type (e.g. Skempton, 1986; Cetin et al., 2004). Soil-pile shaft angle δ_ϕ is often assumed to be the same as soil friction angle for piles cast in situ and about 2/3 of soil friction angle for prefabricated (concrete/steel) driven piles.

- N_q – **ground bearing capacity factor** is defined by many authors as shown in Fig. 7.6 based on the graph by Lambe and Whitman (1979).
- α_p – **ground cohesion mobilization factor along pile shaft** can be obtained from Fig. 7.7, which is based on Tomlinson (2001).

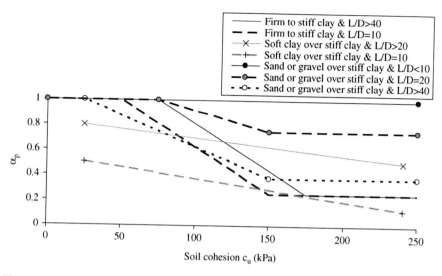

Fig. 7.7   Ground cohesion mobilization factor α along pile shaft

- c_u – **soil cohesion** for OCR = 1 can be determined from Skempton (1957) who suggested a correlation between the undrained shear strength for normally consolidated natural clay and the overburden pressure in one cycle as:

$$\frac{c_{u1}}{\sigma_v'} = 0.11 + 0.0037 \cdot PI, \tag{7.11}$$

where c_{u1} is undrained cohesion in static condition, σ_v' is effective overburden pressure, PI is soil plasticity index in percents. For OCR>1, Ladd and Foot (1974), among others, found that the undrained shear strength of over consolidated clay in static condition is approximately proportional to:

$$\frac{c_{u1}}{\sigma_v'} \cdot OCR^{0.8}, \tag{7.12}$$

where the ratio $c_{u1}\,(\sigma_v')^{-1}$ is given in Equation (7.11). Soil shear strength in cyclic condition depends on number of cycles, cyclic stress or strain amplitude and for fine grained soil on rate of shear in addition to PI of cohesive soil and σ_v' in non-cohesive soil. Soil shear strength in cyclic condition could be determined using laboratory tests such as simple shear described by Finn (1985), for example. Lee and Focht (1976) showed for clay and Boulanger and Idriss (2007) for silt and clay that the soil shear strength in cyclic condition decreases to about a half of undrained cohesion in static condition but sometimes even lower. Figure 7.8 indicates the upper and lower boundaries for clay and the ratios for silt according to the reported results by various authors.

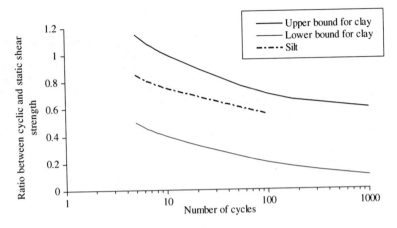

Fig. 7.8 Effect of number of cycles on fine grained soil

The ratios greater than one in Fig. 7.8 (i.e. increased soil shear strength in cyclic condition with relation to the strength in static condition) are caused by rate of shear effect. Parathiras (1995), among others, tested London clay, with plasticity index $PI = 49\%$ and clay content of 60%, and Cowden till, with plasticity index $PI = 21\%$ and clay content of 31%, in a ring shear apparatus. Parathiras (1995) reported an increase of the residual friction angle of $6°$ at 50 kPa compressive stress and $3.5°$ at 400 kPa compressive stress for both soil when the rate of displacement increased from 0.2 to 1.5 cm/s. The increase of the residual friction angle at 50 kPa stress was 48.4% for London clay and 15.2% for Cowden till while at 400 kPa stress was 62.5% for London clay and 10.5% for Cowden till. Rate of shear effects are also noticeable in cone penetration tests. Figure 7.9 shows the ratios between the cone factors (e.g. Lunne et al., 2001) at 2 cm/s cone penetration rate and the factor of nine used in static condition (Equations 7.7 and 7.8).

Pile shaft and toe displacements necessary for estimation of energy at the source of driven piles can be obtained from pile drivability analyses as pile penetration depth per hammer blow. However, not only that such analyse require the use of proprietary software but often they show a rather large scatter of the results. Besides that, it is usually the case that ground resistance to pile penetration is the greatest near pile refusal (end of penetration) and, therefore, the refusal criterion i.e. number of hammer blows per pile penetration over a specified depth can be used for calculation of pile displacement per hammer blow. An example of pile refusal criteria is provided by API RP 2A-WSD (2000). *"Pile driving refusal with a properly operating hammer is defined as the point where pile driving resistance exceeds either 300 blows per foot (0.3 m) for five consecutive feet (1.5 m) or 800 blows per foot (0.3 m) of penetration (This definition applies when the weight of the pile does not exceed four times the weight of the hammer ram. If the pile weight exceeds this, the above blow counts are increased proportionally, but in no case shall they exceed 800 blows for six inches (152 mm) of penetration). If there has been a delay in pile*

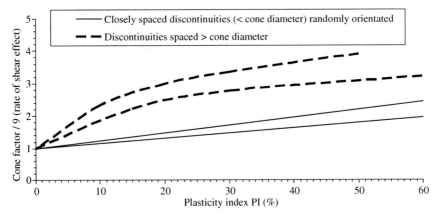

Fig. 7.9 Rate of shear effect (multiplication factor of cohesion in static condition) from cone penetration factor

driving for one hour or longer, the refusal criteria stated above shall not apply until the pile has been advanced at least one foot (0.3 m) following the resumption of pile driving. However, in no case shall the blow count exceed 800 blows for six inches (152 mm) of penetration". Therefore, pile head displacement near the end of pile driving is either 1 mm/blow over 1.5 m penetration or 0.375 mm/blow over 0.3 m penetration and not smaller than 0.19 mm/blow over 0.15 m of penetration. Taking into account pile shortening along its length and above small pile head displacements it follows that the source energy E_o released at pile toe near the end of pile driving is small in comparison with the energy released along the pile shaft.

7.2.1.2 Case Study of Determination of the Peak Particle Velocities During Driving of a Steel H Section Pile by an Impact Hammer

Hiller and Crabb (2000) provided sufficient data concerning the ground vibration induced by driving of piles at the A47 Church road interchange in the United Kingdom. Data provided involve:

- Ground conditions (layering, description of ground types (cohesive/non-cohesive, and N_{SPT} blow count with depth for assessment of soil friction angle ϕ, i.e. soil-shaft friction angle δ_ϕ, and soil cohesion c_u).
- Hammer properties (type and rated energy per blow).
- Recorded hammer blow count versus depth for calculation of pile displacement per blow.
- Peak particle velocities at the ground surface along horizontal distances from the piles for comparison with the results of simple analyses.

Ground water level is assumed to be close to ground surface. The 28 m long steel piles with the cross section of H shape UBP 305×305×186, which is 320.5 mm wide in one and 328.3 mm in the other direction, were driven using Banut 700 type hammer, with nominal energy of 24.52 kJ (corresponding to 5 t ram mass drop from 0.5 m height). Hiller and Crabb (2000) estimated that the driving energy remained constant for a depth range 4 to 28 m apart from the top 4 m because lower driving energy was used for the toeing-in operation. The recorded resultant peak particle velocities at the ground surface with the horizontal distance from the piles are shown in Fig. 7.10 for greater depths of the pile penetrations, which caused the greatest ground vibrations at greater distances. The change in rate of ground vibration attenuation was evident at a horizontal distance of 20 m. Details of the calculations using Equation (2.11) are provided in Section 4 of Appendix. The results of the calculations are shown in Fig. 7.10.

While recorded and calculated peak particle velocities agree rather well at the horizontal distances greater than 20 m, the calculated values are much larger at the horizontal distances up to 10 m from the piles. As already volumetric geometric damping (cube root scaling) is used in the calculations, the large differences at the distances up to 10 m from the pile could be caused by the differences in assumed and actual material damping. Necessary damping coefficients and the results obtained are given in Table 7.1.

The damping coefficients that are necessary to be used in Equation (2.11) to fit the recorded PPV shown in Table 7.1 are not realistic. Most likely cause of recorded peak particle velocities smaller than expected ones at distances to 20 m from the piles was the ground vibration induced built up of excessive pore water pressure and softening of the layer of loose silt, which acted as a base isolator of ground vibration propagation towards the top layer of firm silty clay and the ground surface. Soil

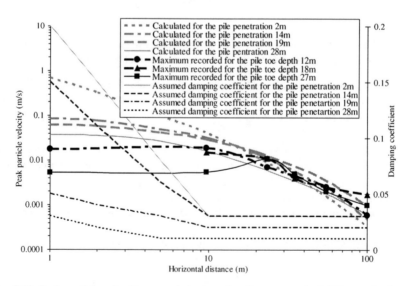

Fig. 7.10 Peak particle velocities recorded when the pile toes reached different depths and calculated for different pile penetration depths in the case study in Section 7.2.1.2

Table 7.1 Inferred damping coefficients at near field distances from the piles necessary to match the recorded and calculated peak particle velocities in the case study in Section 7.2.1.2

Assumed damping coefficient	PPV (m/s) at 1 m horizontal distance	Assumed damping coefficient	PPV (m/s) at 2 m horizontal distance	Assumed damping coefficient	PPV (m/s) at 5 m horizontal distance	Assumed damping coefficient	PPV (m/s) at 10 m horizontal distance
0.99	0.0202	0.85	0.0170	0.4	0.0180	0.1	0.0193
0.70	0.0048	0.60	0.0056	0.4	0.0056	0.15	0.0062

softening does not affect so much velocity of propagation of longitudinal waves but rather velocity of propagation of transversal waves, which decreases to zero through water i.e. liquefied sand. This is not an isolated case as evident from the following case study.

7.2.1.3 Case Study of Determination of the Peak Particle Velocities During Driving of Tubular Steel Piles by Vibratory and Impact Hammer

A number of steel tubular piles with 1.8 m external diameter, 20 mm wall thickness and 34 m length were driven through 4.5 m of sand hydraulically filled over 9.5 m of alluvial sand and gravel resting on top of upper chalk during extension of a port in the United Kingdom. Measured cone penetrometer resistances through upper strata are shown in Fig. 7.11.

The cone penetrometer resistances correspond to sand densities from loose to dense but medium dense on average. The hydraulic filling caused a flow type slope failure and this raised concerned of further slope failures during pile driving by PV 105 M vibratory hammer to about 9 to 11.5 m depth and by IHC S280 hydro hammer for greater depths through chalk. The vibratory hammer operated at the frequency

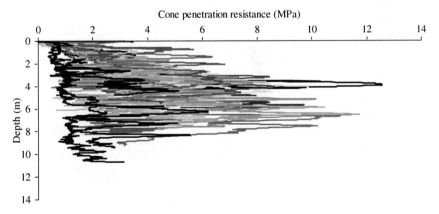

Fig. 7.11 Cone penetration resistances of hydraulically filled sand over alluvial sand and gravel for the case study in Section 7.2.1.3

of 22.5 Hz while the hydro hammer achieved 1.1 to 3.1 strokes per second, which corresponds to a frequency of between 0.32 to 0.91 Hz. The hydro hammer was capable of achieving up to 280 kJ of energy input per blow (corresponding to 13.6 t ram mass and 2.1 m drop). However, the hammer was used at about 1/3 to 1/2 the maximum energy and sometimes as low as 8%.

The flow type failures caused by liquefaction of water saturated sand during earthquakes and rapid filling are not uncommon in loose fine sand of nearly uniform grading as shown in Fig. 7.12 for the sand fill used at the site. The boundary curves in Fig. 7.12 for liquefaction potential are from the report by Japanese Ministry of Transport (1999).

To monitor the situation during the pile driving, a series of geophones were placed perpendicular to the row of piles being driven, as indicated in Fig. 7.13 for some of the piles.

The recorded peak particle velocities versus the horizontal distances are shown in Figs. 7.14 and 7.15.

From Figs. 7.14 and 7.15 follows that both vibratory and hydro hammer, which has not operated at full energy, caused similar peak particle velocities at the ground surface. It is also evident that recorded peak particle velocities have not change much at the horizontal distances up to about 20 m. This can be explained by build up of excessive pore water pressure and soil softening or liquefaction due to ground vibration caused by pile driving. The argument becomes more convincing when the recorded peak ground accelerations caused by the pile driving versus the horizontal distances are shown in Figs. 7.16 and 7.17.

From Figs. 7.16 and 7.17 follows that the pattern of the recorded peak ground acceleration change with distance follows the pattern of the recorded peak ground velocities. Peak ground accelerations in the range from 0.06 to 0.42 g for loose to dense sand with less than 5% fines represent border values of the acceleration between liquefied and non-liquefied condition according to the chart by Seed et al. (1985). The recorded peak ground accelerations in excess of the border values indicate that sand at the location did not liquefy but only softened.

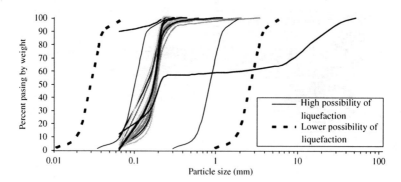

Fig. 7.12 Particle size distributions by weight for the hydraulically filled sand in the case study in Section 7.2.1.3

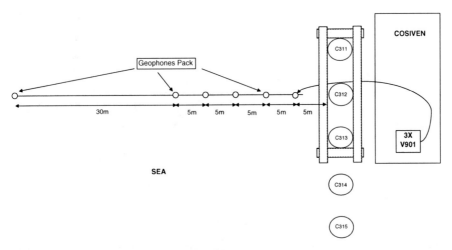

Fig. 7.13 Example layout for the monitoring of the ground vibration caused by pile driving described in the case study in Section 7.2.1.3

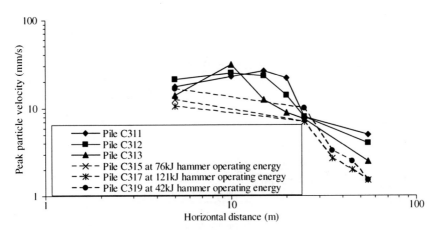

Fig. 7.14 Recorded maximum peak particle velocities versus horizontal distances from operation of the hydro hammer IHC S280 in the case study in Section 7.2.1.3

Vibratory hammers induce harmonic motion and hence a product of peak particle velocity, 2π and their vibration frequency should be equal to the peak ground acceleration, i.e. from Fig. 7.15, $1 - 45$ mm/s $* 2\pi * 22.5$ Hz/9810 $= 0.014 - 0.65$ g, which is within the range of measured peak ground accelerations from 0.014 to 3 g shown in Fig. 7.17.

The recorded blow count of the hydro hammer in upper chalk was in the range from 5 to 40 per 0.25 m pile penetration, with an average at about 15 per 0.25 m pile penetration, which corresponds to the lower bound of predicted blow counts. The cumulative number of hammer blow counts (cycles) is over 700. Details of the

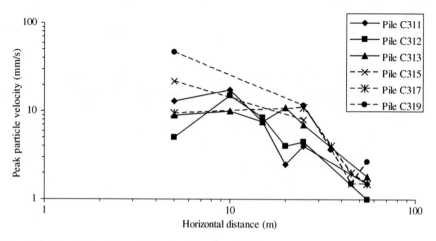

Fig. 7.15 Recorded peak particle velocities versus horizontal distances from operation of the vibratory hammer PV 105 M in the case study in Section 7.2.1.3

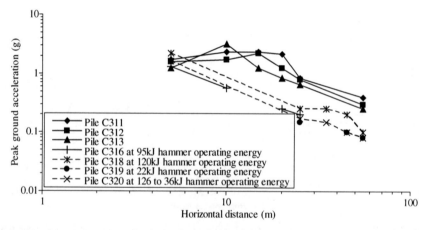

Fig. 7.16 Recorded maximum peak ground accelerations versus horizontal distances from operation of the hydro hammer IHC S280 in the case study in Section 7.2.1.3

calculations are given in Section 4 of Appendix. The results of the calculations are shown in Fig. 7.18.

The results shown in Fig. 7.18 are influenced by assumed ground damping particularly at vibration source to site distances greater than 10 m. Accuracy of ground damping at greater distances between vibration source and a site is important only if calculated peak particle velocities are of significance concerning sensitivities of recipients (Section 1.2). Ground damping depends on ground strain, which in turn depends on both particle and wave propagation velocity (Equations 2.2 and 2.3) so

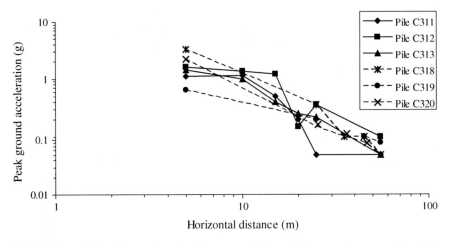

Fig. 7.17 Recorded peak ground accelerations versus horizontal distances from operation of the vibratory hammer PV 105 M in the case study in Section 7.2.1.3

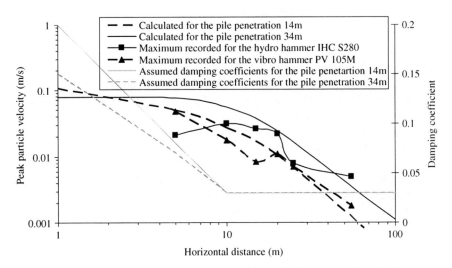

Fig. 7.18 Calculated peak particle velocities versus horizontal distances at different pile penetration depths and maximum recorded PPV for the two hammers in the case study in Section 7.2.1.3

that a recursive relationship exist between particle velocity and ground damping. Wave propagation velocity is also strain dependent.

7.2.1.4 Case Study of Determination of the Peak Particle Velocities During Driving of Tubular and Sheet Piles by Vibratory Hammers

Hiller and Crab (2000) provided results of recorded ground peak particle velocities caused by vibratory hammers and detailed data for two locations. The 900 mm

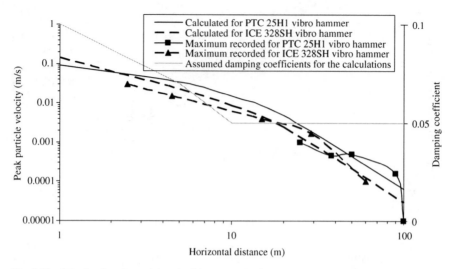

Fig. 7.19 Calculated peak particle velocities versus horizontal distances and maximum recorded PPV for the two hammers in the case study in Section 7.2.1.4

diameter and 14 m long casing at A11 Foxes Bridge was driven by a vibratory hammer type PTC 25H1 (with energy per cycle of 4.97 kJ and frequency of 29 Hz) through 3 m of very dense sand overlying chalk of variable grade. The 5 m long steel sheet L40 was driven by a vibratory hammer type ICE 328SH (with energy per cycle of 0.99/1.17 kJ and frequency of 46.7 Hz) through colliery waste. Details of the calculations are given in Section 4 of Appendix while the results are shown in Fig. 7.19.

The results shown in Fig. 7.19 depend on assumed soil damping particularly at the distances greater than 20 m. Providing that soil damping is accurate then the simplified method (Equation 2.11) can provide accurate results.

7.2.2 Soil Shallow Compaction

A brief description of the problem is provided in Section 1.3.1.2. Some recorded peak particle velocities caused by weight drop during dynamic compaction and by vibratory equipment are shown in Figs. 7.20 and 7.21.

Recorded peak particle velocities shown in Fig. 7.20 vary in a wide range because of the variations of the source energies and therefore their prediction using simplified analyses may be necessary. Recorded peak particle velocities in Fig. 7.21 caused by vibratory compaction equipments indicate that the peak particle velocities can exceed the limited values specified in codes (e.g. Fig. 1.2) only if ground compaction is performed very close to structures, in which case simplified methods or field measurements could be used for assessment of ground vibration amplitudes.

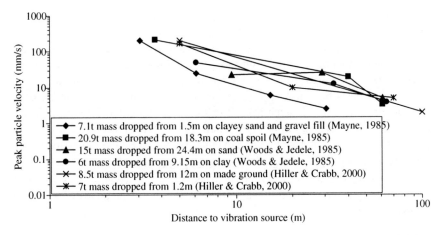

Fig. 7.20   Recorded peak particle velocities versus distance to the masses dropped from heights

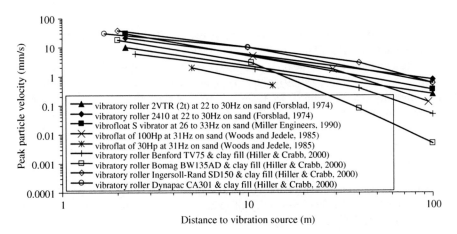

Fig. 7.21   Recorded peak particle velocities versus distances caused by vibratory compaction equipments

Besides the upper bounds of recorded peak particle velocities shown in Fig. 7.20, empirical relationships exist for peak particle velocity attenuation with distance from the source. For example, Hiller and Crabb (2000) carried out linear regression analysis on data from different compaction passes from both the controlled trial and from construction sites and obtained the following expression for the resultant particle velocity (mm/s) that is obtained from three componential velocities (also BS 5228-2, 2009)

$$v_{res} = k_s \cdot \sqrt{n_d} \cdot \left(\frac{A_r}{x_r + w_d} \right)^{1.5}, \tag{7.13}$$

where the coefficient $k_s =75$ for an average value i.e. a 50% probability of the vibration level being exceeded, $=143$ for a 33% probability of the vibration level being exceeded and $=276$ for a 5% probability of the vibration level being exceeded, $n_d \leq 2$ is the number of vibration drums, A_r is the nominal amplitude of the vibrating roller (mm) in the range from 0.4 to 1.7 mm, x_r is the distance along the ground surface from the roller (m) in the range from 2 to 110 m, w_d is the width of the vibrating drum (m) in the range from 0.75 to 2.2 m. Equation (7.13) is applicable for a travel speed of approximately 2 km/h. For significantly different operating speeds of rollers, v_{res} in Equation (7.13) could be scaled by the ratio between $2^{1/2}$ and (roller speed in km/h)$^{1/2}$ according to Hiller and Crabb (2000).

For dynamic ground compaction by dropping heavy weights, Mayne (1985), for example, presented the following formula for the resultant particle velocity (mm/s):

$$v_{res} \leq 92 \cdot \left(\frac{\sqrt{M_d \cdot H_d}}{x_i} \right)^{1.7}, \qquad (7.14)$$

where M_d is the tamper mass (tonnes), H_d is the drop height (m), x_i is the distance from impact (m). BS 5228-2 (2009) recommends to use 0.037 multiplying coefficient instead of 92 but for the product $M_d H_d$ expressed in J instead of tm like Mayne (1985). Also BS 5228-2 (2009) limits the values of x_i in the range from 5 to 100 m. For vibrating stone columns, BS 5228-2 (2009) provides the following formula for the resultant particle velocity:

$$v_{res} = \frac{k_c}{x^{1.4}}, \qquad (7.15)$$

where $k_c = 33$ (for 50% probability of exceedance), 44 (for 33.3% probability of exceedance), 95 (for 5% probability of exceedance), x is the horizontal distance range from 8 to 100 m.

7.2.2.1   Case Study of Determination of the Peak Particle Velocities During Installation of Stone Columns by a Vibratory Probe

The hydraulically filled sand has been compacted at the location that is mentioned in the case study in Section 7.2.1.3 using a vibratory probe, which penetrates ground under self-weight, and by installation of stone columns. Horizontal vibrations of the probe are induced by rotating eccentric weights mounted on a shaft driven by a motor housed within the casing, Fig. 7.22. The probe displaces and densifies sand along its penetration depth. Stone backfill is introduced in controlled batches, either from the surface down the annulus created by penetration of the probe (Fig. 7.23) or through feeder tubes directed to the tip of the probe. Re-penetration of each stone backfill batch forces the stone radially into the surrounding soil, forming a vibratory stone column that is tightly interlocked with the soil in a system which

Fig. 7.22 Vibro probe used for installation of stone columns at the location in case study in Section 7.2.2.1

Fig. 7.23 Stone backfill about to be poured around a vibratory probe to create a stone column. The bars with flags mark positions of the stone columns yet to be placed

has lower compressibility, higher shear strength and permeability than natural soil (e.g. Schaefer et al., 1997). Despite extensive research and development concerning stone columns, field trials are still the best and most reliable method for the column design.

The recorded peak particle velocities during the installation of a stone column at the location and extrapolated values when it is assumed that excess pore water pressure and soil softening does not cause soil liquefaction are shown in Fig. 7.23. For estimated vibration energy of the probe of $50 \times 1.5 \times 3 \times 0.45 \times 0.025 \sim 2.5$ kJ at a depth of 2.5 m (where 50 kPa is assumed effective overburden stress at assumed penetration depth of 5 m of the probe, 1.5 is estimated coefficient of lateral soil pressure, 3 m is estimated length of the probe, 0.45 m is estimated width of the probe, 0.025 m is estimated horizontal movement of the probe), the calculated peak

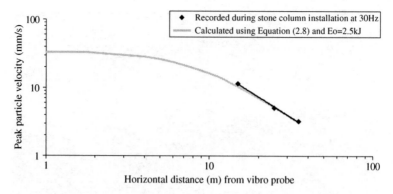

Fig. 7.24 Recorded and calculated peak particle velocities due to installation of a stone column in the case study in Section 7.2.2.1

ground velocities from Equation (2.8), for assumed damping coefficient of 0.01 are shown in Fig. 7.24.

In reality, vibratory probe causes sand liquefaction, which in turn would cause much less ground velocities at the surface in vicinity of the vibratory probe, perhaps not more than about 10 mm/s, based on the recorded 'clipped' peak particle velocities at the horizontal distances up to about 20 m in the case studies in Sections 7.2.1.2 and 7.2.1.3.

7.2.2.2 Case Study of Determination of the Peak Particle Velocities During Fill Compaction by Vibratory Rollers

Hiller and Crabb (2000) reported on the results of measurements of ground vibration caused by fill compaction using vibratory rollers during construction and from a test site. They also commented on a number of factors that affect ground vibration induced by vibratory rollers such as:

- **Centrifugal force and frequency.** Hiller and Crabb (2000) stated that *"preliminary analysis had indicated that the centrifugal force is not an appropriate parameter for predicting vibration. Since rollers are designed to operate at frequencies above the resonance, the frequency per se appears to be unlikely to make a significant contribution to the resulting level of vibration. However, if the operating frequency were to coincide with the characteristic frequency of the soil, then problems might be expected"*. The resonance can happen during start and stop of vibratory roller operation.
- **Static linear load.** A linear relation between the static linear load and the resultant peak particle velocity is suggested from an upper bound envelope to the data.
- **Nominal amplitude.** There was clearly an increase in the resultant peak particle velocity with increasing amplitude of vibration.

- **Travel speed.** From tests carried out using one vibratory roller over a range of speeds between zero and 6.7 km/h, it was found that the resultant peak particle velocity was approximately related to the inverse of the square root of the travel speed.
- **Number of drums.** The particle velocity arising from a double drum roller was approximately proportional to $2^{1/2}$.
- **Energy transmitted into the fill.** A linear regression analyses provided the following relationship for the resultant peak particle velocity

$$v_o = 2.07 \cdot \sqrt{A_r \cdot L_s \cdot w_d \cdot g} - 1.4, \qquad (7.16)$$

where A_r is the nominal drum amplitude, L_s is the static linear load, w_d is the drum width, g is the gravitational acceleration, the square root expression is in J when A_r, L_s, w_d, g are in SI units and v_o is in mm/s. Equation (7.16) does not account for the travel speed or the number of vibrating drums and is applicable at a distance of 2 m from a roller operating on clay.

- **Fill properties.** Hiller and Crabb (2000) stated that *"The first pass of any item of plant always gave rise to lower levels of vibration than subsequent passes. Vibration from rollers operating on the clay was greater than when operating on the hogging in all cases except for the Bomag Variomatic".* (Hogging is sandy gravel.) Also, larger vibration arising from plant operating on less stiff fill in majority of cases although contradictory evidence was observed. Hiller and Crabb (2000) stated that *"Comparing the vibration levels with the stiffness for all materials showed no correlation between the stiffness and the peak particle velocity. For individual materials, however, there was some evidence that the particle velocity was higher when the stiffness was higher for a particular material".*

The test fill was placed in a trench having fine sand at its bottom at a depth of 1.5 m. Well graded granular fill was used for the first three layers across the whole base of the excavation. First two layers of the fill were compacted using five passes of a smooth drum tandem roller because of the presence of loose sand underneath. The third final layer of the fill was compacted with two passes of a vibratory roller and two passes using the roller as a dead weight. Above the three initial layers, the further 13 test layers from excavated London Clay and sandy gravel were placed until the compacted fill level had reached original ground level. Vibration measurements were made for the final compaction pass of each test layer. Triaxial arrays of geophones were positioned at distances of 1, 4, 10, 40 and 100 m from the edge of the test fill. Details of four among other vibratory rollers used at the test and construction sites are given in Table 7.2 from Hiller and Crabb (2000).

Details of the calculations performed using Equation (2.18), for assumed Poisson's ratio of 0.35 and transversal wave velocity of 130 m/s i.e. shear modulus G of 30 MPa are given in Section 5 of Appendix while the results are shown

Table 7.2 Summary data for some of the plants used in the controlled trials in the case study in Section 7.2.2.2

| Model | Type | Drum width (m) | Mass (kg/m) | | High setting | | | Low setting | | |
			Front	Rear	Amplitude (mm)	Frequency (Hz)	Centrifugal force (kN)	Amplitude (mm)	Frequency (Hz)	Centrifugal force (kN)
Benford TV75	Tandem	0.75	920	920	0.5	50	9.8	–	–	–
Bomag BW135AD	Tandem	1.3	1330	1390	0.4	60	37	0.4	50	26
Ingersoll-Rand SD150	Single drum	2.14	4367	–	1.77	26.5	245	0.89	26.5	123
Dynapac CA301	Single drum	2.13	3150	–	1.72	30	249	0.84	33	146

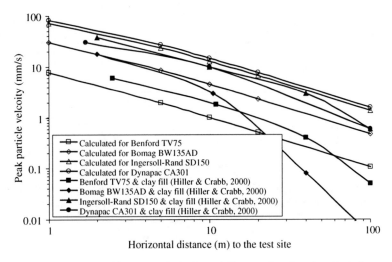

Fig. 7.25 Peak particle velocities versus the horizontal distances based on the calculations and the recordings in the case study in Section 7.2.2.2

in Fig. 7.25 together with the upper bounds of the recorded peak particle velocities from the controlled trials.

From Fig. 7.25, it is evident that the agreement between the calculated and recorded peak particle velocities is better at the source to site distances up to about 10 m, where it matters more because the velocities are larger. Differences between the calculated and recorded values for the source to site distances up to about 10 m are most likely caused by assumed soil properties (Poisson's ratio and shear modulus). Increased attenuation of peak ground velocities at the distances greater than about 10 m could be caused by the presence of soil with greater stiffness than the soil present at the test field as the amplitudes are inversely proportional to soil stiffness according to Equation (2.18).

7.2.3 Demolition of Structures

The problem is introduced in Section 1.3.1.3. Some available attenuation relationships from recorded peak particle velocities caused by demolition of structures are shown in Figs. 7.26 and 7.27.

In Fig. 7.27, the peak particle velocities above 20 mm/s at the horizontal distances greater than 50 m are obtained from the demolition of the buildings in Mexico City, which is located partly over soft lake deposits with shear modulus near the ground surface of only about 1.5 MPa. Excluding the peak particle velocities from Mexico City and from demolition of Thornhill cooling tower in Fig. 7.26, the remaining peak particle velocities indicate similar attenuation relationships from the demolition of the cooling towers and the buildings. Large scatter in Figs. 7.26 and 7.27

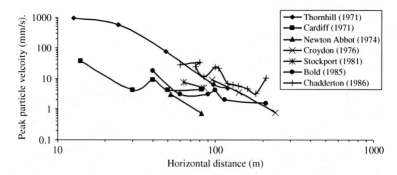

Fig. 7.26 Recorded peak particle velocities at different horizontal distances during demolition of cooling towers at different locations and in different years (based on data from Eldred and Skipp, 1998)

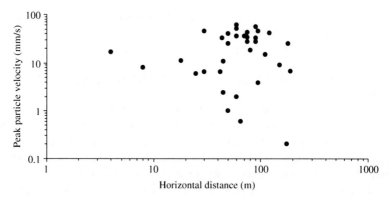

Fig. 7.27 Recorded peak particle velocities at different horizontal distances during demolition of buildings (based on data from Dowding, 2000)

is caused by different impact energies and ground conditions and, hence, use of simplified methods could increase accuracy of prediction.

7.2.3.1 Case Study of Determination of the Peak Particle Velocities During Demolition of a Cooling Tower at Thornhill in 1971

Eldred and Skipp (1998) provided data concerning demolition of a cooling tower at Thornhill in 1971. The tower height was 88 m and its weight was 50 MN. The base diameter was 57.1 m, the waste diameter 31.7 m and the cornice diameter 38.1 m. A sketched view of the tower is shown in Fig. 7.28.

The tower footing was founded on medium dense sand with gravel which exists under about 1.3 m thick layer of ash. Demolition is performed by blasting tower legs on one side of its base so that the initial impact energy was $0.5 \times 50\,\mathrm{MN} \times 5.5$

Fig. 7.28   Sketched view of
the Thornhill tower based on
the decryption by Eldred and
Skipp (1998)

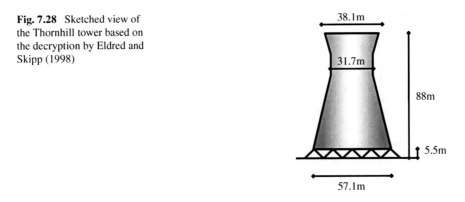

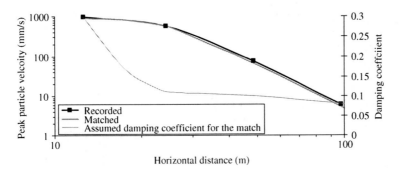

Fig. 7.29   Calculated and recorded peak particle velocities in the case study in Section 7.2.3.1

m = 137.5 MNm. The crashing of the rest of the cylindrical shell that followed
could have lasted more than $(2 \times 88 \times 9.81^{-1})^{1/2} = 4.2$ s if it was a free fall. Peak
particle velocities are calculated using an equation similar to Equation (2.8) but for
a half space on which the impact energy was applied $(0.5 \times 4/3r^3\pi)$. The results are
shown in Fig. 7.29.

Damping coefficients shown in Fig. 7.29 are obtained by matching the calculated and recorded peak particle velocities. Once again, it has been demonstrated
that knowledge of ground damping coefficient is very important for accuracy of
prediction of peak particle velocities particularly at greater distances. As ground
damping coefficient is dependent on strain, which in turn is dependent on the ratio
between peak particle velocity and wave propagation velocity (which is also dependent on strain), an iterative calculation is necessary until assumed and calculated
peak particle velocities are almost identical. In this case, ground damping coefficient
and ground wave propagation velocity dependence on strain were not provided by
Eldred and Skipp (1998).

7.2.4 Blasting in Construction and Mining Industries

The problem is introduced in Section 1.3.1.4. Some available empirical relationships of peak particle attenuation with distance from blasting are shown in Figs. 7.30, 7.31, and 7.32.

From Fig. 7.30 it is evident that even if scaled distances, with explosive mass in kg, are used the scatter in the peak particle velocities is large. Therefore, simplified analyses could be used to increase accuracy of prediction of PPV. The peak particle velocities at the blast location to receiver distances of up to about 25 m, which are reported by Kahriman (2004), exceed the upper bound values according to Caltrans (2001) in Fig. 7.32 and deserve further attention. Different peak particle

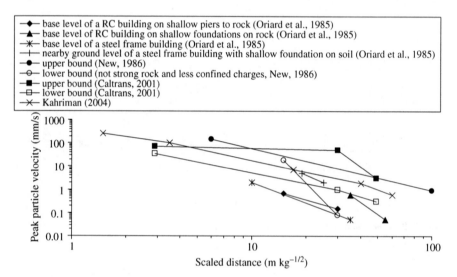

Fig. 7.30 Peak particle velocities versus scaled distances m/kg$^{1/2}$

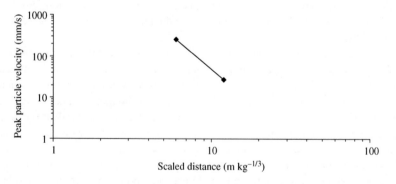

Fig. 7.31 Peak particle velocities versus scaled distances m/kg$^{1/3}$ (List et al., 1985)

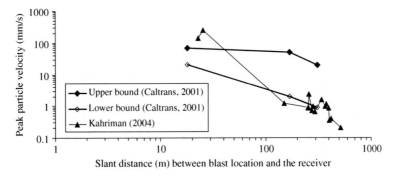

Fig. 7.32 Peak particle velocities versus slant distances

velocity attenuation relationships exist in literature. For example, U.S. Bureau of Mines (1971) established an upper limit of ground vibrations caused by blasting as:

$$PPV = 714 \cdot \left(\frac{\sqrt{W_e}}{D_b} \right)^{1.6}, \tag{7.17}$$

where PPV is peak particle velocity (mm/s), D_b is distance (m) to blast location, W_e mass (kg) of explosive used. Peak particle velocities predicted by Equation (7.17) are situated roughly in the middle between the upper and lower bound limits by Caltrans (2001), shown in Figs. 7.30 and 7.32.

7.2.4.1 Case Study of Determination of Peak Particle Velocities Caused by Bench Blasting at a Limestone Quarry

Kahriman (2004) provided results of measurements of peak particle velocities from 73 blast events during the bench blast optimization studies in a limestone quarry near Istanbul in Turkey. The heights of the 1st and 3rd benches were 20 m and of the 2nd and 4th benches 30 m. The holes for explosive charges were vertical and 105 mm in diameter. Three rows of holes for explosive charges per bench were used. The row spacing was 2 m for the 1st and 3rd bench and 2.5 m for the 2nd and 4th bench. The spacing of the holes per row was 2.5 m for the 1st and 3rd bench and 3 m for the 2nd and 4th bench. The first row of the holes was placed 3.5 m (burden thickness) from the slope for the 1st and 3rd bench and 4 m for the 2nd and 4th bench. Explosive length within the holes was 15 m of which 1 m was located below the bench level (sub drilling) with a plug (stemming) of 6 m for the 1st and 3rd bench and 24 m of which 2 m was located below the bench level (sub drilling) with a plug (stemming) of 8 m for the 2nd and 4th bench. The blasting operation used 42 ms delay between rows and a 17 ms delay between holes within a row with 25 ms interval used for inner borehole detonation. The explosive used was ANFO + 5% Al (blasting agent), Rovex 650 and gelatine dynamite (priming). ANFO's detonation velocity is about

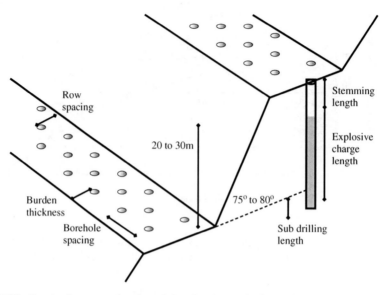

Fig. 7.33 Sketch of cross section through benches in a rock slope with the holes for explosive charges in the case study in Section 7.2.4.1

Table 7.3 Test data in the case study in Section 7.2.4.1

No	Charge per delay (kg)	Frequency (Hz)	Distance (m)	Peak particle velocity (mm/s)
1	180	20	23	144
2	242	45	25.6	250
3	69	10	150	1.2
4	175	9.4	254	0.9
5	81	11	255	0.85
6	538	10	257	2.35
7	97	91	275	0.7
8	206	5.6	283	0.95
9	93	67	295	0.65
10	165	17	340	1.55
11	83	16	368	0.95
12	242	9.1	376	1.15
13	354	9.6	400	0.85
14	180	41	403	0.35
15	180	20	418	0.4
16	170	9	520	0.2

4200 m/s according to various sources. The energy released was proportional to the product of a half of the mass of explosive and the detonation velocity squared. A sketch of the benches with holes for explosive charges is shown in Fig. 7.33 while the test data are given in Table 7.3.

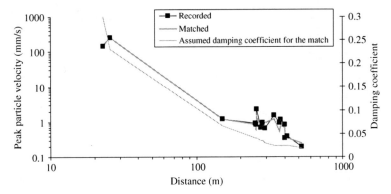

Fig. 7.34 Recorded and matched peak particle velocities versus distances for assumed damping coefficients in the case study in Section 7.2.4.1

The results of use of Equation (2.8) for calculation of PPV (with assumed ρ of 2500 kg/m^3 and damping coefficients) are shown in Fig. 7.34.

From Fig. 7.34 it follows that the simple method can provide a good prediction of the peak particle velocities caused by the mine blasting providing that ground damping coefficient variation with strain is known and used.

7.2.4.2 Case Study of Determination of Peak Particle Velocities Caused by Blasting for a Pipeline Installation

The installation of twin 1 m diameter pipelines using explosives in rock caused complaints from the owners of houses located at approximately 120 m nearest distance from the blasting operations in Caucasus. There is no available information on ground conditions at the locations of the houses. A number of boreholes that are drilled along the pipeline route indicated presence of a layer of stiff clay from about 0.5 m to more than 3 m deep overlying basalt or dolerite. Occasionally, thick layer of tuff (consolidated volcanic ash ejected from vents during a volcanic eruption) is encountered in the boreholes. Only one recorded peak particle velocity at a distance of 70 m from the pipeline route exists at the location of the houses. The recorded peak particle velocities versus distances scaled with the square root of the ratio between 9.3 kg per delay of the explosive used at the location of the houses (big dot) and other locations along the pipeline route (dashes) are shown in Fig. 7.35 together with the vibration frequencies in Hz.

Unfortunately, the number of recorded PPV at distances larger than 20 m of interest is limited for determination of a reliable PPV attenuation relationship with distance. As repetition of blasting at the pipeline location would damage the pipeline, a weight dropping trial was arranged at the location in order to try to fill the gap in data at distances greater than 20 m. A 700 kg steel mass lifted 4 m above ground was used as shown in Fig. 7.36. The energy used was $700 \times 9.81 \times 4/1000 = 27.47$ kJ assuming free fall of the weight without loss due to the cable friction.

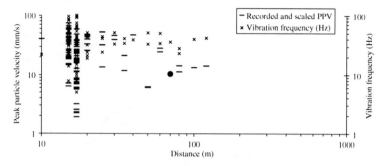

Fig. 7.35 Recorded and scaled peak particle velocities versus distances from the pipeline route in the case study in Section 7.2.4.2

Fig. 7.36 Weight dropping trial in the case study in Section 7.2.4.2

Total of five locations between the pipeline route and the houses were tested and the ground acceleration monitored at five distances from the weight drop using the triaxial accelerometers fastened to a spike shown in Fig. 7.37. Distances to the five locations from the pipeline route were 40, 85, 85, 135, and 135 m respectively.

The peak particle velocities obtained after integration in time of the recorded accelerations are shown in Fig. 7.38. The frequency of ground vibration caused by the weight drop was from 40 to 50 Hz at the distances greater than 20 m, i.e. similar to the frequency range of ground vibration caused by the blasting. Under impulse load, ground oscillates at its fundamental frequency of vibration, which can be inferred from Fourier spectra (Section 4.4.1).

Equation (2.8) is used for the calculations in both the weight drop and the blasting case, for which it is assumed that the detonation velocity was 5000 m/s. However for the weight drop case, only a half of radiation damping is considered ($0.5 \times 4/3\ r^3\pi$) because of the weight drop on a half space. Inferred damping coefficients for the weight drop case are used for the blasting case. As the calculated PPV from

Fig. 7.37   The accelerometer arrangement used in the case study in Section 7.2.4.2

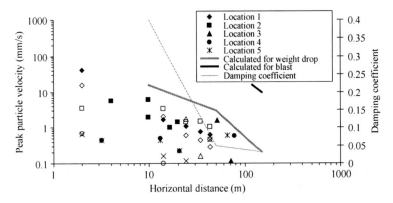

Fig. 7.38   Peak particle velocities versus distances recorded from the weight drop (in the horizontal direction – filled symbols, in the vertical direction – no fill) and calculated for the weight drop and the blasting in the case study in Section 7.2.4.2

the blasting are larger than from the weight drop, the actual damping coefficients could only be larger in the blasting case and therefore the results obtained are on a conservative side.

Based on the graph DIN residential & similar buildings containing people in Fig. 1.2 and calculated PPV range from 10 to 16 mm/s in the frequency range from 40 to 50 Hz, the PPV due to the blasting at the distances greater than 120 m from the pipeline route was probably just below the graph values, which indicate minimum peak particle velocity necessary to cause so called cosmetic damage. However, the PPV due to the blasting at the distances greater than 120 m from the pipeline route was probably just above the values in the graph DIN vulnerable to vibration &

important buildings in Fig. 1.2. The peak particle velocities greater than 0.1 mm/s in the frequency range from 40 to 50 Hz are noticeable by people based on Fig. 1.1

7.2.5 Soil Deep Compaction by Explosives

The problem is introduced in Section 1.3.1.5. The method has great cost effectiveness (e.g. Gohl et al., 2000) but its effectiveness depends on a number of factors. The peak particle velocities versus scale distances reported in literature are shown in Fig. 7.39. The difference is caused by the different factors affecting the peak particle velocities.

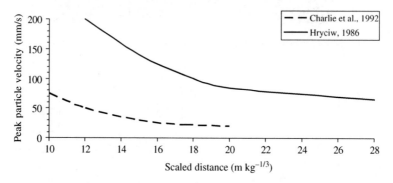

Fig. 7.39 Peak particle velocities versus scaled distances from blast densification of sand

7.2.5.1 Case Study of Determination of Peak Particle Velocities Caused by Densification of Pond Ash by Blasting

Gandhi et al. (1999) reported on the results of a series of test blasting carried to assess the effectiveness of blasting to increase density of pond ash in India. Ninety explosions comprising 15 single blasts, with varying depths and quantities of charges, and 3 group blasts, each comprising of 25 charges placed at various spacing, were carried out. Gandhi et al. (1999) concluded that *deep blasting may be an effective technique for modest compaction of loose fly ash deposit. Blasting has been found to be simple, easy and cost effective and it does not require special construction machinery. At the time of testing, the pond was filled with a 12 m thick ash deposit, and the ground water table was 0.8–4 m below the surface.* As ash was transported hydraulically as slurry, and because of low density of fly ash, the average unit density was only 1500 kg/m^3. A slurry type explosive (a mixture of ammonium nitrate and sodium nitrate with aluminium powder and a sensitizing agent) was used. Each charge had 83 mm diameter and 500 mm length and weighted 2.78 kg; 2–6 cartridges were used per borehole. Assumed detonation velocity is 5000 m/s. The data for the single blasts are given in Table 7.4. In two cases out of 15, peak particle

Table 7.4 Data for the single blasts in the case study in Section 7.2.5.1

Blast No	3	4	5	7	8	9	10	2	11	12	13	14	15
Charge weight per delay (kg)	11.1	8.3	5.5	13.9	11.1	8.3	5.5	27.8	16.7	13.9	11.1	8.3	5.5
Depth of charge (m)	6	6	6	8	8	8	8	9	9	9	9	9	9
Peak particle velocity (mm/s) at 30 m distance	12	8.6	2.2	9	7	10	6	20	12		8.5	8.5	7.4	5.5

acceleration was measured instead of peak particle velocity. For 16.7 kg explosive mass at 6 and 8 m average depths, the recorded peak acceleration was 1.6 m/s^2. The piezoelectric sensor, operating in the frequency range from 0.3 to 15 kHz, were oriented in the vertical direction and placed on a thick metal plate on the ground surface.

Using Equation (2.8) and assumed damping coefficient of 0.3 at 30 m distance from the blasts, calculated peak particle velocities are in a good agreement with the maximum recorded PPV as shown in Fig. 7.40.

The peak particle velocity of 16 mm/s at a distance of 30 m from the outer blast point was measured for the group blast III only due to non availability of the measuring instrument for the group blast I and II. The total charge mass of 83.4 kg was placed at 9 m average depth in five boreholes. In the layout, the borehole locations formed the tips of three adjacent equilateral triangles with side length of 36 m. The total length of the set up was 72 m and the width 31 m. Using Equation (2.12) and assumed damping coefficient of 0.3 at 30 m distance, the calculated peak particle velocity is 2.3 mm/s. Using Equation (2.13) with $D_s = 3$ m in Fig. 2.2c and assumed damping coefficient of 0.3 at 30 m distance, the calculated peak particle velocity is 10.4 mm/s. A prismatic vibration source model (Equation 2.13) provides better prediction of the PPV than a planar vibration source model (Equation 2.12) in the case considered. The difference between recorded and calculated PPV could be caused by assumed ground damping coefficient.

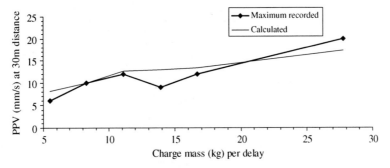

Fig. 7.40 Maximum recorded and calculated peak particle velocities versus explosive charge mass in the case study in Section 7.2.5.1

7.3 Vibration Caused by Trains and Road Vehicles

7.3.1 Train Caused Vibration

The problem is introduced in Section 1.3.2.1. Some available recorded peak particle velocities due to high speed trains are shown in Fig. 7.41. The E1 to E3 sites in Fig. 7.41 represent different locations at Site E, which contains medium stiff to stiff sandy clay and medium dense clayey sands under the ground surface. The ground water level depth is greater than 12 m. The range of vibration frequencies at Site E was mainly from about 8 to 25 Hz (extremely from 6 to 40 Hz).

The variation in recorded peak particle velocities is significant for different cases and it is interesting to analyse some of them using simplified methods. Bahrekazemi (2004) reviewed the state of the art concerning ground vibration caused by trains. Among available attenuation relationships listed by Bahrekazemi (2004), the atten-uation relationship by the U.S. department of Transport (U.S.-D.O.T., 1995, 1998) considers the effect of a large number of influential factors on the root mean square (r.m.s.) particle velocity. The effect of train speed on r.m.s. particle velocity is shown in Fig. 7.42.

The multiplication factors of the r.m.s. particle velocities for the 240 km/h reference train speed are as follows (Bahrekazemi, 2004).

For the source factors:

- 2.0 for 480 km/h train speed, 1.33 for 320 km/h train speed, 1.0 for 240 km/h train speed, 0.67 for 160 km/h train speed, 0.50 for 120 km/h train speed
- 3.16 for worn wheels or wheels with flat parts, worn or corrugated crack, crossovers and other special track work
- 0.18 for floating slab track bed

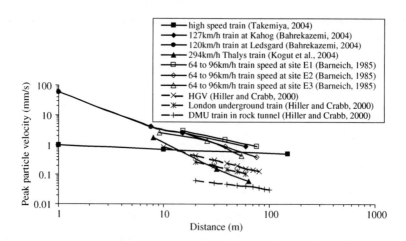

Fig. 7.41 Peak particle velocities versus distances to train rails

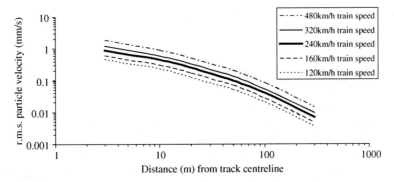

Fig. 7.42 Root mean square (r.m.s.) particle velocities versus distances (based on U.S.-D.O.T., 1998)

- 0.32 for ballast mats
- 0.56 for high resilience fasteners
- 0.32 for resiliently supported ties
- 0.32 for Ariel/viaduct structure
- 1.0 for open cut
- 0.56 for station
- 0.71 for cut and cover tunnel
- 0.18 for rock based track

For the path factors:

- 3.16 for efficient propagation in soil
- 1.26 at 15 m distance, 1.58 at 30 m distance, 2.0 at 45 m distance and 2.82 at 60 m distance for wave propagation through rock
- 0.56 for a wood frame
- 0.45 for 1 to 2 story commercial building
- 0.32 for 2 to 4 story masonry
- 0.32 for large masonry on piles
- 0.22 for large masonry on spread footings
- 1.0 for foundation on rock

For the receiver factors:

- 0.8 for 1 to 5 floors above grade
- 0.9 for 5 to 10 floors above grade
- 2.0 due to resonances of floors, walls, and ceilings

Clearly, the coefficient greater than 1 indicates worse condition than the referent case and less than 1 better condition than the referent case conditions. The adjustments for wheel and rail condition are not cumulative. When more than one adjustment may apply, the general rule is to apply only the largest adjustment.

7.3.1.1 Case Study of Determination of Peak Particle Velocities Caused by High Speed Thalys Train

Kogut et al. (2004) presented experimental validation of a numerical model for high speed train induced vibrations on the line L2 between Brussels and Koln. The site near high speed train track in Lincent contains a layer of about 8 m thick sandy clay above a sand stratum. The spectral analyses of surface waves (the method described in Section 6.2.3) and the seismic cone penetration tests (the method described in Section 6.3.3) indicated the transversal wave velocity of about 150 m/s in the top 1.5 m with the velocity rate increase with depth of about 50 m/s per metre to 5.5 m depth, at which the results of the measurements are available.

Prior to the tests of vibrations cause by trains, the transfer functions between the track and the free field have been measured for several impacts of a falling weight on the rail head. At 8 m distance from the track, the frequency content of the free field response is broad with a maximum at about 30 Hz, while at 64 m distance from the track the frequency content becomes narrower and the maximum shifts to about 20 Hz, with an average value of about 25 Hz.

The time history of the recorded velocity of the sleeper (below the rail) during the passage of the Thalys train with a speed of 294 km/h revealed the peak particle velocity of 40 mm/s. The 11 peaks in the time history correspond to every axle of the train. Train axles are combined in pairs and are called bogies. From the length of the time history of about 2.3 s and the speed of the train of 294 km/h it follows that the train length was about 188 m. The frequency content of the velocity of the sleeper during the passage of the Thalys train with a speed of 294 km/h indicate a quasi discrete spectrum with peaks at the fundamental bogie passage frequency of 4.37 Hz (for bogie spacing of 18.7 m) and its higher harmonics corresponding to the axle passage frequency of 27.22 Hz (for axle spacing of 3.0 m). The product between the sleeper velocity of 40 mm/s and the period of its vibration due to the axle passage of $27.22^{-1} = 0.0367$ s represents the sleeper settlement of 1.47 mm on the axle passage.

Shortly before the measurements of the ground velocities, the railway company had used a vehicle to measure the track unevenness. From the frequency content of the interaction force at the first axle computed from the unevenness data by Kogut et al. (2004) it follows that the axle force is 136.1 kN for the axle passage frequency of 27.22 Hz. The product of the force of 136.1 kN and the sleeper settlement of 1.47 mm represent the energy applied on the ground at each axle of 0.200 kJ.

Equation (2.18) is used for the calculation of peak particle velocities for the averaged vibration frequency of 25 Hz, assumed Poisson's ratio of 0.45, the shear modulus $G = 1800 \times 200^2 \times 0.001 = 72000$ kPa and the force of $136.1 \times 2 = 272.2$ kN for two axle forces spaced at 3.0 m distance. The results are shown in Fig. 7.43.

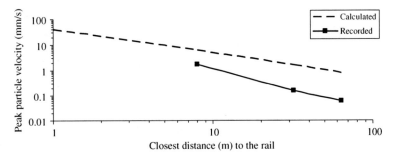

Fig. 7.43 Calculated and recorded peak particle velocities versus distances in the case study in Section 7.3.1.1

Although the calculated peak particle velocity at 1 m distance to the rail is similar to the recorded PPV of the sleeper of 40 mm/s, the calculated PPV at greater distances are greater than the recorded PPV. One possible reason for the discrepancy could be high variability of the ground condition across the site with stiffer ground existing at larger distances.

7.3.1.2   Case Study of Determination of Peak Particle Velocities Caused by High Speed Train at Kahog in Sweden

Bahrekazemi (2004) presented the results of measurements and analyses of data for several test sites including Kahog site, which is located about 10 km north of Gothenburg in Sweden. About 1.5 m thick crust of organic soft clay at the site overlays more than 15 m thick layer of silty clay with some thin layers of silt and sand. Soil unit density near the top is about 1600 kg/m^3 and increases to 1900 kg/m^3 at 14 m depth. The ground water depth is about 2 m below the ground surface. The transversal wave velocity at the site is about 70 m/s near the top and increases with depth.

The wheel force measured in the rail during passage of X2000 train at 127 km/h speed varied in the range from 60 to 90 kN. The total duration of the record is about 4.6 s, which correspond to the train length of about 162 m. The spikes in the record are arranged in pairs with the spacing between them of 0.071 s, which indicates the spacing of 2.5 m between the axles of a bogie, and with the spacing between the spike pairs of about 0.426 s, which indicates the spacing of 15 m between the adjacent axles of two bogies of a carriage. The frequency of vibration caused by passage of a pair of axles of a bogie over a sleeper is $0.071^{-1} = 14.18$ Hz.

Equation (2.18) is use for the calculation of peak particle velocities for the vibration frequency of 14 Hz, assumed Poisson's ratio of 0.45, the shear modulus $G = 1600 \times 100^2 \times 0.001 = 16000$ kPa and dynamic force of $75 \times 2 = 150$ kN for two axle forces spaced at 2.5 m distance. The results are shown in Fig. 7.44.

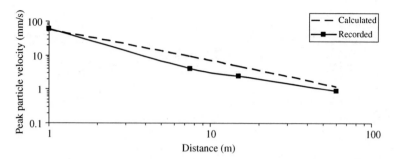

Fig. 7.44 Calculated and recorded peak particle velocities versus distances in the case study in Section 7.3.1.2

7.3.2 Vehicle Caused Vibration

The problem is introduced in Section 1.3.2.2. Ground vibration induced by passage of vehicles is unlikely to cause structural damage (even so called cosmetic cracking) but could exceed the perception level for humans and therefore trigger complaints to a local authority and/or an environmental agency. For this reason it is necessary to be able to estimate peak particle velocity caused by vehicles. Summary of road traffic induced peak particle velocities at distances from 3 to 6 m from vehicles is shown in Fig. 7.45, based on data by Barneich (1985). The vibration frequency range was mainly from about 10 to 20 Hz (with extremes from 3 to 35 Hz).

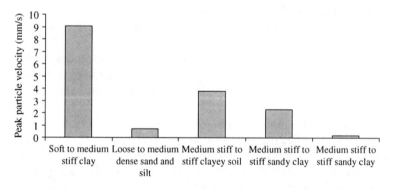

Fig. 7.45 Peak particle velocities induced by road traffic at 3 to 6m distances from vehicles

7.3.2.1 An Example of Calculation of Peak Particle Velocity Caused by a Wheel Drop into a Road Hole

Let suppose that a passenger vehicle applies 2.5 kN load per wheel and that its wheel drops into a 0.1 m deep pot hole in a poorly maintained local road. The wheel falling into the hole will apply energy of $2.5 \times 0.1 = 0.25$ kJ on the ground.

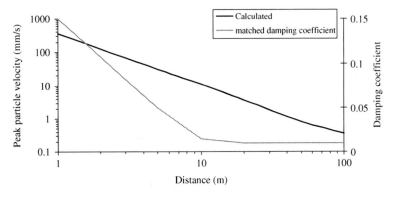

Fig. 7.46 Peak particle velocities versus distances in the example in Section 7.3.2.1

Equation (2.8) but for a half space radiation damping through a half sphere with volume $2/3r^3\pi$ will be used to calculate expected peak particle velocities (PPV) at different distances from the hole for assumed ground unit density of 1800 kg/m^3 and transversal wave velocity of 180 m/s. As PPV depends on ground damping coefficient, which depends on strain level, which depends on the ratio between PPV and wave propagation velocity it follows that a recursive relationship exist. The iterative calculation is performed in steps until the calculated and initially assumed PPV are accurate enough. Only one step was sufficient to determine the damping coefficient and final PPV shown in Fig. 7.46.

7.4 Machinery Caused Vibration

The problem is introduced in Section 1.3.3. Various codes and standards exist for foundations of machinery. For example, CP 2012-1 (1974) is intended for design and construction of foundations for reciprocating machinery with rotating frequency in the range from 5 to 25 Hz. The recommendations are generally applicable only to the design of foundations that are represented by an undamped single mass spring system. DIN 4024-1 (1988) specifies requirements for steel or reinforced concrete foundations that support machinery with mainly rotating elements. DIN 4024-2 (1991) specifies requirements for rigid machine foundation blocks made from steel or reinforced concrete that are intended to support machinery subjected to periodic vibration.

7.4.1 Industrial Hammers Caused Vibration

Svinkin (2002) presented results of the measurements of foundation and ground displacements caused by impact of forge hammer and falling weights. He also proposed a method, which is applicable at distances greater than about 0.5–4 of the

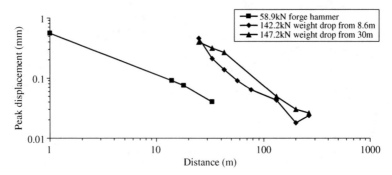

Fig. 7.47 Variation of peak displacements with distances from a forge hammer and weight drops (based on data by Svinkin, 2002)

dominant wavelength, for predicting the complete time domain vibration records of soil and structures prior to installation of foundations for impact machines. The recorded foundation and ground displacements are shown in Fig. 7.47.

7.4.1.1 Case Study of Determination of Peak Particle Velocities Caused by Weight Drops

Svinkin (2002) has not specified the drop height of the forge hammer and there-fore only the cases of weight drops will be considered. The peak particle velocities are calculated from the recorded time histories of ground displacements as the first derivatives in time, Fig. 7.48.

The weight drops were performed over about 1.6 m thick layer of loose sand, underlain by about 6.8 m thick layer of medium dense sand, over 1 m thick layer of sandy clay and about 10 m thick layer of sand. The ground water level was about 6 m deep below the ground surface. Equation (2.8) but for a half space radiation damping through a half sphere with volume $2/3r^3\pi$ is used for prediction of peak particle velocities. The calculated PPV with assumed damping coefficients are shown in Fig. 7.48.

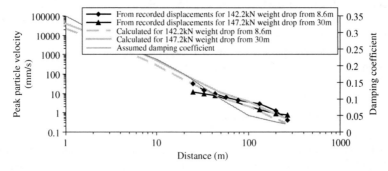

Fig. 7.48 Calculated peak particle velocities versus distances in the case study in Section 7.4.1.1

The calculated peak particle velocities in vicinity of an impact hammer are rather large and are likely to cause use of measures to minimize them, which are considered in Chapter 8.

7.4.2 Case Study of Determination of Ground Vibration Caused by a Compressor

It was required to provide foundation for a large compressor in Holland. A sketch of the compressor's view is shown in Fig. 7.49 together with persons standing on the top and next to it for scale.

The total mass of the rotor is 24 t and of the compressor 67 t. Other relevant data are shown in Table 7.5.

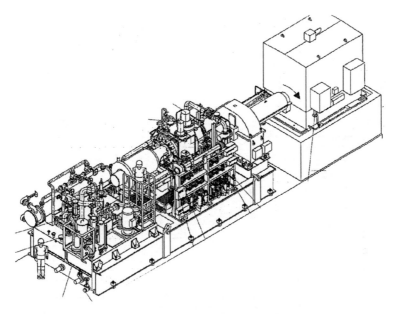

Fig. 7.49 A sketch of the view to a compressor in Holland in the case study in Section 7.4.2

Table 7.5 Data for the components in the case study in Section 7.4.2

Component	Circular frequency (cycles/min)	Unbalanced mass (kg)	Incidental unbalanced force (N) = six times the operational force
Driver rotor	1495	7100	41650
Wheel shaft	1495	4630	27160
Pinion shaft	11440	210	9430
Compressor rotor	11440	375	16835

The requirements of manufacturer of the compressor are:

- Under accidental loading, the upper surface of the machine foundation shall have peak velocity of less than 6 mm/s.
- During the operation, the upper surface of the machine foundation shall have peak velocity of less than 2.8 mm/s.
- The natural frequency of the foundation is outside the range 12.5±20% to 25±20% Hz of the frequency of the rotor.
- The natural frequency of the foundation is outside the range 95+20% to 191+20% Hz of the frequency of the compressor.

The reinforced concrete foundation dimensions in mm are shown in Fig. 7.50.

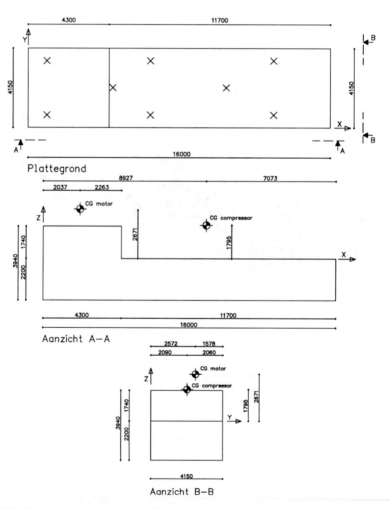

Fig. 7.50 Dimension of the foundation block in mm in the case study in Section 7.4.2

The site investigation involved cone penetration tests with some of them including measurements of side friction in addition to the cone tip resistance. Based on the diagram by Robertson et al. (1986), corrected cone resistance and the friction ratio ranges, it follows that the investigated soil is sand to silty sand. Rix and Stokoe (1992) suggested the following formula for the maximum shear modulus G_{max} of uncemented quartz sand.

$$\left(\frac{G_{max}}{q_c}\right)_{average} = 1634 \cdot \left(\frac{q_c}{\sqrt{\sigma'_{vo}}}\right)^{-0.75},$$

$$Range = Average \pm \frac{Average}{2}$$

(7.18)

where q_c – cone tip resistance and σ'_{vo} – effective overburden pressure are in kPa. Transversal wave velocity v_t is calculated from a simple expression, $v_t = (G_{max} \rho^{-1})^{1/2}$ where ρ is soil unit density. For assumed $\rho = 2000$ kg/m^3 and q_c range recorded at the site, transversal wave velocity versus level is shown in Fig. 7.51.

A reviewer of the original design recommended not only a change of the pile group under the foundation block of the compressor but also use of rubber bearings under the block. Use of rubber bearings requires their checking and replacement so that it was necessary to assess this design solution independently. The assessment involved both numerical modelling of continua using 2-D dynamic PLAXIS and QUAD4M software and simplified analyses, of which only the later are presented here. First, the vibration of the foundation block without piles is considered using the discrete element model by Wolf (1994), sketched in Fig. 7.52.

Input data and details of the calculations are given in Section 6 of Appendix. The results are shown in Fig. 7.53.

The calculated peak velocity of the shallow foundation is 3.3 mm/s, the frequency of free horizontal vibration of the foundation is 6.56 Hz, the frequency of free rotational vibration of the foundation is 7.55 Hz, and therefore satisfactory. The peak particle velocity of 3.3 mm/s corresponds to strain of about $3.3 \times 150000^{-1} = 2 \times 10^{-5}$ (according to Equation (2.3) for the transversal wave velocity near ground

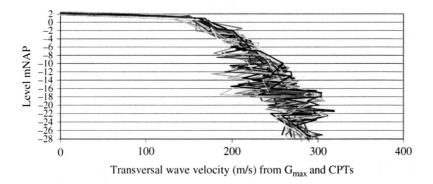

Fig. 7.51 Inferred transversal wave velocity versus level in the case study in Section 7.4.2

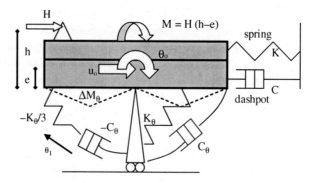

Fig. 7.52 Discrete element model of a shallow foundation for the case study in Section 7.4.2

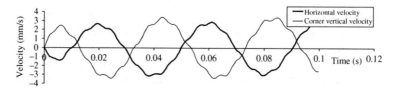

Fig. 7.53 Results of computation for the shallow foundation in the case study in Section 7.4.2

surface of 150000 mm/s), which is considered a small strain and therefore G_{max} is applicable without its reduction for strain level . However, the top few meters of sandy soil is in a loose to medium dense state (transversal wave velocity less than 180 m/s), which is capable of developing of large settlements under cyclic loading and of excess pore water pressure. Therefore, either the layer density needs to be increase by deep compaction or piled foundation that transfer loads to greater depth needs to be used. The increased layer density would result in the standard penetration test blow count of up to 50 and the transversal wave velocity of up to $80 \times 50^{0.333} = 295$ m/s (Equation 6.1). The calculated peak velocity of the shallow foundation would be 1.8 mm/s, the frequency of free horizontal vibration of the foundation would be 12.9 Hz, the frequency of free rotational vibration of the foundation would be 14.9 Hz, and therefore inside the range (10–15 Hz) to be avoided according to the manufacturer's criteria. The option with piled foundation is considered next.

The locations of piles in the original design are shown as crosses in Fig. 7.50. The piles are 0.45 m in diameter and 10 m long. The pile group i.e. the equivalent spring stiffness and frequencies of free vibrations are analysed using CONAN software (Section 5.3) for a stack of equivalent disks over the pile depth of 10 m. The radii of the equivalent disks, which are calculated according to Equations (5.5), (5.6), and (5.7), are given in Table 7.6.

Assumed damping coefficient is 0.05. Calculated dynamic stiffness coefficient S_e is shown in Fig. 7.54, combining the static stiffness coefficient K_e with spring $k(a_o)$ and damping coefficient $c(a_o)$ according to formula (Wolf and Deeks, 2004):

Table 7.6 Radii (m) of the equivalent disks in the case study in Section 7.4.2

Motion	Vertical	Horizontal	Rotating around longer dimension
Pile group	0.64	0.37	1.29
Foundation block	4.60	3.32	3.32

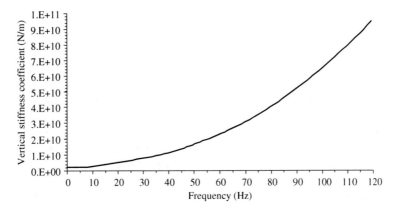

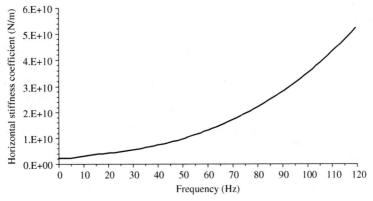

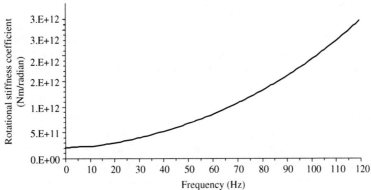

Fig. 7.54 Equivalent spring stiffness for the case study in Section 7.4.2

Table 7.7 Foundation velocities (mm/s) in the case study in Section 7.4.2

Motion	Vertical	Horizontal	Rotating around longer dimension
Peak velocity (mm/s)	1.6	2.2	0.2

$$S_e = K_e \cdot \sqrt{k(a_o)^2 + c(a_o)^2 \cdot a_o^2} \, , \tag{7.19}$$

where $a_o = 2\pi\, f\, r_e v_t^{-1}$ is dimensionless frequency, f is frequency, r_e is radius of the equivalent disk, $v_t = 150$ m/s is the transversal wave velocity of the top soil layer.

For incidental unbalanced dynamic forces i.e. rotating moments acting on the driver rotor and wheel shaft at the frequency of 25 Hz, the peak foundation velocities calculated from the ratio between the dynamic force/moment and the dynamic stiffness coefficient at the frequency of 25 Hz times the circular frequency $(25 \times 2\pi)$, times a half of the block width $= 2.075$ m for the rotational motion are given in Table 7.7.

The peak foundation velocity is smaller than 6.0 mm/s as required so it is satisfactory. The calculated ratios between the foundation and ground vibration amplitudes are shown in Fig. 7.55 combining the calculated real and imaginary parts from the CONAN results as $(\text{real}^2 + \text{imaginary}^2)^{1/2}$ (Wolf, 1994)

The frequencies (>0) of the foundation vibration at the fundamental and higher modes can be inferred from the peaks in the graphs in Fig. 7.55. The peak in Fig. 7.55 for the horizontal direction is within the range of frequencies that must be avoided (20–30 Hz) according to the manufacturer's criteria and therefore the piled foundation is not satisfactory. A simplified consideration of the vibration of concrete block on rubber bearings placed within a concrete bin on piles is given in Section 8.2.1.1.

7.4.3 Case Study of Determination of Ground Vibration Caused by a Gas Turbine

It was required to check the vibration properties of the foundation under a gas turbine with accompanying compressors and driven packages in Gabon. The approximate centres of gravities of two compressors, two driven package lifts, and the gas turbine are shown in Fig. 7.56.

The reinforced concrete raft foundation width is 4.1 m, length 13.0 m and thickness 0.5 m. The foundation and machine mass is 101 t. The mass eccentricity (e, in Fig. 7.52) is 1.06 m. Other input data are given in Table 7.8 for the shut down case of the rotors and Section 7 of Appendix. The horizontal and vertical load act simultaneously. The vertical loads on opposite sides of the machines are coupled so that when one acts upwards the other acts downwards and vice versa.

The machinery manufacturer requirements are:

- The componential velocity amplitude at the location of the machine bearing housing does not exceed 2 mm/s.
- The peak to peak amplitude of any part of the foundation is less than 0.05 mm.

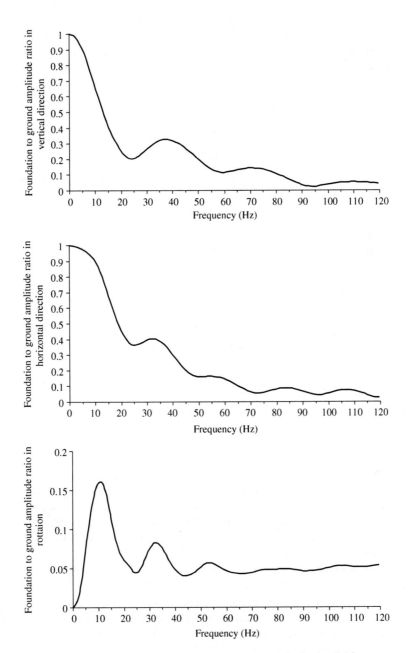

Fig. 7.55 Foundation to ground amplitude ratios in the case study in Section 7.4.2

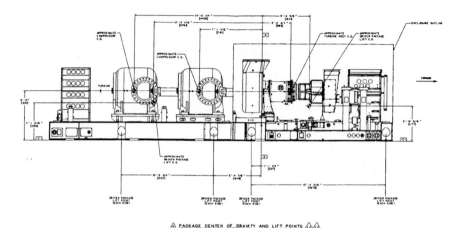

Fig. 7.56 Side view of the locations of centres of gravities of two compressors, two driven package lifts and a gas turbine (the total length 12.0 m)

Table 7.8 Dynamic loads and their frequencies in the case study in Section 7.4.3

Load no (direction: h – horizontal, v – vertical) Property	1(h)	2(h)	3(v)	4(h)	5(v)	6(h)	7(v)
Horizontal unbalanced force (N)	1148	3265	2980	5004	7073	5004	7073
Eccentricity (m) from the total mass centre of gravity	1.09	1.09	2*2.33/2	1.09	2*2.33/2	1.09	2*2.33/2
Rotational frequency (radians/s)	1571	1571	1571	1497	1497	1497	1497

The foundation is placed over in situ or compacted coarse granular laterite soil (weathered rock). The standard penetration test blow count varies in the range from 12 to over 30 at shallower depths under the foundation. The transversal wave velocity varies in the range from $80 \times 12^{0.333} = 180$ m/s to over $80 \times 30^{0.333} = 250$ m/s based in Equation (6.1). Details of the computation are given in Section 7 of Appendix and the results shown in Figs. 7.57 and 7.58.

Calculated maximum foundation velocity is 1.2 mm/s (<2 mm/s) and peak to peak amplitude of any part of the foundation is 0.0015 mm (<0.05 mm) so that the foundation satisfies the manufacturer's criteria. Calculated frequency of the free horizontal foundation vibration varies in the range from 17 to 24 Hz and of the rotational vibration from 31 to 43 Hz, which are much smaller than the vibration frequency of the machinery of about 240 Hz.

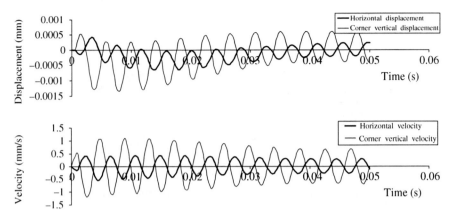

Fig. 7.57 Results of the calculation for the transversal wave velocity of 180 m/s in the case study in Section 7.4.3

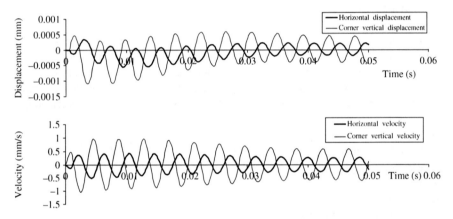

Fig. 7.58 Results of the calculation for the transversal wave velocity of 250 m/s in the case study in Section 7.4.3

7.4.4 Tunnel Boring Machines Caused Vibration

Hiller and Crab (2000) summarised data on resultant peak particle velocities various distances from various tunnel boring machines (TBM) and tools, Fig. 7.59.

The peak particle velocities can be calculated using Equation (2.8). The source energy is proportional to the energy of tunnel excavating machine times an efficiency factor, which is machine dependent. BS 5228-2 (2009) provides the following equation for the resultant particle velocity due to tunnelling:

$$v_{res} \leq \frac{180}{r^{1.3}} , \qquad (7.20)$$

where the slant distance r range is from 10 to 100 m.

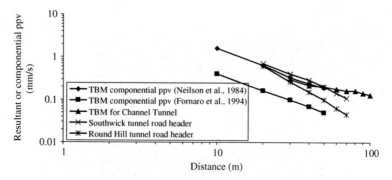

Fig. 7.59 Resultant or componential peak particle velocities (ppv) versus distances from tunnelling machines (based on Hiller and Crabb, 2000)

7.5 Summary

A number of case studies that are considered in this chapter indicate that the simple equations from Sections 2.3 can successfully predict peak ground velocities from different vibration sources providing that ground properties such as damping coefficient/shear stiffness variation with shear strain are known and used in the prediction. Likely ranges of variation of damping coefficients with distances are indicated for:

- Impact and vibratory pile drivers
- Demolition of structures by explosives
- Blasting of rock
- Mass drop from height for soil testing
- Soil compaction by explosives
- Vehicle wheel drop into road holes
- Industrial hammers

The use of ground shear stiffness is demonstrated for:

- Ground compaction by vibratory rollers
- Train passing

Chapter 8
Control of Ground and Foundation Vibration

8.1 Introduction

When predicted peak particle velocities or frequency contents in Chapter 7 exceed limits set by relevant standard or equipment manufacturer (e.g. Section 1.2) then ground (and foundation) vibration needs to be minimized if possible, its spread through ground minimised or the sensitive receiver provided with a protection against vibration. The simplified approach considered in this volume is based on the use of Equation (2.8), which indicates that ground vibration is minimized by increase in both mass (density ρ) and material damping ($e^{k \cdot r}$), and Equation (2.18), which indicates that ground vibration is minimized by decreasing source frequency f, and by increasing Poisson's ratio v and shear stiffness modulus G.

The objective of this chapter is to describe the ways of minimization of source vibration, of minimisation of its spread through ground and protection of receivers from ground vibration by different means and their simplified analyses.

8.2 Minimization at Source

This is preferable option of control of ground vibration as it prevents spread of vibration into environment and protects all potential receivers from the effects of vibrations.

8.2.1 Base Isolation

Soong and Dargush (1997), for example, listed the following types of modern isolation devices:

- Elastomeric bearings
- Lead rubber bearings
- Sliding friction pendulum

M. Srbulov, *Ground Vibration Engineering*, Geotechnical, Geological, and Earthquake Engineering 12, DOI 10.1007/978-90-481-9082-9_8, © Springer Science+Business Media B.V. 2010

The function of isolators.

The isolators are used to decrease the frequency and increase damping of bases of vibration sources and, hence, decrease peak particle velocities and/or change the frequency content of vibration. The bearings are formed by alternate vulcanized elastomer/rubber layers and steel plates, bonded together by use of chemical compounds. LESSLOSS (2007), for example, states that the sliding isolation pendulum consists of a pivoting steel lens with a slider, which slides along a stainless steel spherically curved sliding surface (the concave plate). More details about this type of base isolator as well as other innovative anti-seismic systems with materials and dimensions, specifications, calculation examples, testing and quality control, installation, inspection and maintenance are provided in LESSLOSS (2007). Elastomer and rubber bearing are subject to ageing and therefore require inspection and may need replacement if their prolonged use is required. In the past, steel springs have been used for base isolation. Rather detailed consideration of vibration isolation systems is provided in specialist books, for example by Rivin (2003).

8.2.1.1 Cases Study of Base Isolation by Rubber Bearings of the Foundation Block of a Compressor

Details of the foundation block with loading from a compressor in Holland are given in Section 7.4.2. The reviewer of the original design proposed the use 10 rubber bearings under the foundation block, with the following properties for each bearing:

- Width and length 0.5 × 0.5 m
- Height 0.275 m
- Shear modulus 0.77 MPa
- Vertical stiffness 52 MN/m
- Horizontal stiffness 0.75 MN/m

The spacing between the centres of the bearings is 2.85 m in the perpendicular direction of the block, no damping coefficient was specified. For the concrete block with machinery, the masses and their locations considered are given in Table 8.1.

For an equivalent concrete block, which is $1.92 \times 2 = 3.84$ m high and 16 m long, the block width is $533826/(3.84 \times 16 \times 2500) = 3.48$ m. A sketch of the equivalent block is shown in Fig. 8.1. The rotor and wheel shaft are located at $2.2 + 2.67 - 1.92 = 2.95$ m above the centre of gravity of the equivalent block.

Table 8.1 Masses and their locations considered in the case study in Section 8.2.1.1

Type	Mass (kg)	Location above block underside (m)	Mass moment (kg m)
Concrete block	365200	2.2/2	401720
Concrete pedestal	77626	2.2 + 1.74/2	238312
Rotor and wheel shaft	24000	2.2 + 2.67	116880
Compressor and pinion shaft	67000	2.2 + 1.795	267665
Summary (average)	533826	(1024577/533826 = 1.92 m)	1024577

Fig. 8.1 A sketch of cross section through the equivalent block in the case study in Section 8.2.1.1

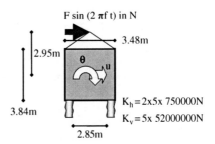

The equation of horizontal motion of the block centre of gravity is:

$$m \cdot \frac{d^2u}{dt^2} + K_h \cdot \left(u - \theta \cdot \frac{3.84}{2}\right) = F \cdot \sin(2\pi \cdot f \cdot t) \quad i.e.$$

$$\frac{d^2u}{dt^2} + \omega^2 \cdot \left(u - \theta \cdot \frac{3.84}{2}\right) = F \cdot \frac{\sin(2\pi \cdot f \cdot t)}{m} , \qquad (8.1)$$

where $\omega^2 = (K_h m^{-1})$, $m = 533826$ kg, $K_h = 7500000$ N, $F = 68810$ N, $f = 25$ Hz.

The closed form solution of the non-homogeneous second order differential Equation (8.1) with constant coefficients is for the under damped case (when damping is less than the critical damping) and without consideration of the block rotation θ (e.g. Kramer, 1996):

$$u = C_1 \cdot \sin(\omega \cdot t) + C_2 \cdot \cos(\omega \cdot t) + \frac{F}{K_h \cdot (1 - \beta^2)} \cdot \sin(2\pi \cdot f \cdot t) , \qquad (8.2)$$

where $\beta = 2\pi f \omega^{-1}$. At $t = 0$, $u = 0$, $C_2 = 0$. At $t = 0$, $du(dt)^{-1} = 0$, $C_1 = -F K_h^{-1} \beta (1-\beta^2)^{-1}$. Without consideration of the block rotation θ, the maximum velocity in the horizontal direction = 3.3 mm/s in Section 8 of Appendix. The frequency of free vibration of the foundation block is in the horizontal direction $(2\pi)^{-1}(7500000 * 533826^{-1})^{1/2} = 0.6$ Hz and in the vertical direction $(2\pi)^{-1}(520000000 * 533826^{-1})^{1/2} = 5.0$ Hz, which is smaller than the upper limit of 10 Hz specified by the manufacturer of the compressor and therefore satisfactory.

The equation of the rotational block motion is:

$$I \cdot \frac{d^2\theta}{dt^2} + K_v \cdot \left(\frac{2.85}{2}\right)^2 \cdot \theta + K_h \cdot \frac{3.84}{2} \cdot \left(\frac{3.84}{2} \cdot \theta - u\right) = 2.95 \cdot F \cdot \sin(2\pi \cdot f \cdot t) , \qquad (8.3)$$

where $I = 12^{-1} \times 533826 \times (3.48^2 + 3.84^2) = 1194435$ kg m^2. After substitution of u from Equation (8.2) into Equation (8.3) taking into account that $C_2 = 0$, $C_1 = -F K_h^{-1} \beta (1-\beta^2)^{-1}$:

$$I \cdot \frac{d^2\theta}{dt^2} + \left[K_v \cdot \left(\frac{2.85}{2} \right)^2 + K_h \cdot \left(\frac{3.84}{2} \right)^2 \right] \cdot \theta = 2.95 \cdot F \cdot \sin(2\pi \cdot f \cdot t)$$

$$\text{(8.4)}$$

$$- \frac{3.84}{2} \cdot \frac{F \cdot \beta}{1 - \beta^2} \cdot \sin(\omega \cdot t) + \frac{3.84}{2} \cdot \frac{F}{1 - \beta^2} \cdot \sin(2\pi \cdot f \cdot t)$$

The closed form solution of the non-homogeneous second order differential equation (8.4) with constant coefficients is:

$$\theta = C_a \cdot \sin(\Omega \cdot t) + C_b \cdot \cos(\Omega \cdot t) + \frac{2.95 \cdot F}{K_\theta \cdot (1 - B^2)} \cdot \sin(2\pi \cdot f \cdot t)$$

$$+ \frac{3.84}{2} \cdot \frac{F}{K_\theta \cdot (1 - \beta^2) \cdot (1 - B^2)} \cdot \sin(2\pi \cdot f \cdot t) \qquad \text{(8.5)}$$

$$- \frac{3.84}{2} \cdot \frac{F \cdot \beta}{K_\theta \cdot (1 - \beta^2) \cdot (1 - \Gamma^2)} \cdot \sin(\omega \cdot t),$$

Where $\Omega^2 = \frac{K_\theta}{I}$, $K_\theta = K_v \cdot \left(\frac{2.85}{2} \right)^2 + K_h \cdot \left(\frac{3.84}{2} \right)^2$, $B = 2\pi f \Omega^{-1}$, $\Gamma = \omega \Omega^{-1}$. For $t = 0$, $\theta = 0$, $C_b = 0$. For $t = 0$, $d\theta (dt)^{-1} = 0$, $C_a = \frac{-2.95 \cdot F \cdot B}{K_\theta \cdot (1 - B^2)} - \frac{3.84}{2} \cdot \frac{F \cdot B}{K_\theta \cdot (1 - \beta^2) \cdot (1 - B^2)} + \frac{3.84}{2} \cdot \frac{F \cdot \beta \cdot \Gamma}{K_\theta \cdot (1 - \beta^2) \cdot (1 - \Gamma^2)}$. The time history of velocity of the block rotation is shown in Fig. 8.2 based on the calculation results in Section 8 of Appendix.

The frequency of free rotational vibration of the foundation block is:
$(2\pi)^{-1} \{ [5 \times 52000000 \times (2.85/2)^2 + 10 \times 750000 \times (3.84/2)^2] / (1194435 + 533826 \times 1.92^2) \}^{1/2} = 2.1$ Hz, which is smaller than the upper limit of 10 Hz specified by the manufacturer of the compressor and therefore satisfactory.

The approximate solution exhibits chaotic behaviour as shown by its phase graph in Fig. 8.3. Baker and Gollub (1990), for example, described several techniques for

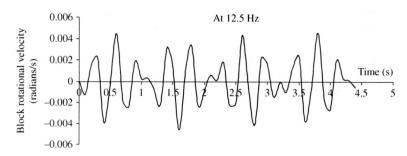

Fig. 8.2 Rotational velocity of the foundation block from the closed form solution in the case study in Section 8.2.1.1

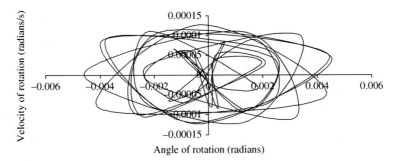

Fig. 8.3 Phase graph of the closed form solution of angle of rotation versus the velocity of rotation of the foundation block on rubber bearings in the case study in Section 8.2.1.1

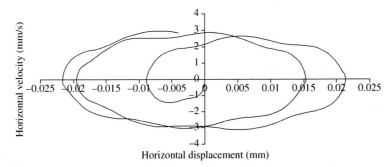

Fig. 8.4 Phase graph of the horizontal displacement versus horizontal velocity of the foundation block placed directly on soil in the case study in Section 7.4.2

investigation of chaotic behaviour of even deterministic systems, which are based on deterministic equations such as those derived from Newton's second law of motion. Necessary conditions for chaotic motion are that the system has at least three independent dynamical variables and that the variables are coupled by non-linear relations. Among available techniques, phase graph, i.e. the plot of system displacement versus its velocity, is used in Fig. 8.3. Crossing of the trajectories of system motion violates uniqueness of trajectories in a deterministic dynamical system and indicates its chaotic behaviour.

The phase graph of vibration of the foundation block directly placed on soil in the case study in Section 7.4.2 is of an elliptical shape, which is typical for a harmonic motion, Fig. 8.4. The graph undulations are caused by higher frequency but lower amplitude motions of the pinion shaft and compressor rotor.

The Equations (8.1) and (8.3) can be expressed in the finite difference form. Details on finite difference approximations can be found in Wood (1990), for example.

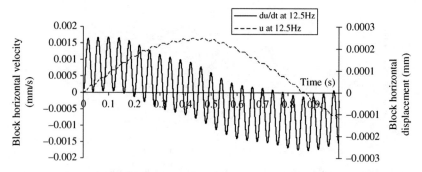

Fig. 8.5 Block horizontal velocity and displacement versus time in the case study in Section 8.2.1.1

$$m \cdot \frac{u_{next} - 2 \cdot u_{current} + u_{previous}}{(\Delta t)^2} + K_h \cdot \left(u_{current} - \theta_{current} \cdot \frac{3.84}{2} \right)$$

$$= F \cdot \sin(2\pi \cdot f \cdot t_{current})$$

$$I \cdot \frac{\theta_{next} - 2 \cdot \theta_{current} + \theta_{previous}}{(\Delta t)^2} + \left(K_v \cdot \left(\frac{2.85}{2} \right)^2 + K_h \cdot \left(\frac{3.84}{2} \right)^2 \right) \cdot \theta_{current}$$

$$= 2.95 \cdot F \cdot \sin(2\pi \cdot f \cdot t_{current}) + K_h \cdot \frac{3.84}{2} \cdot u_{current}$$

$$(8.6)$$

In Equation (8.6), only u_{next} and θ_{next} are unknown values and can be calculated directly from the current and previous values. At $t = 0$, $u_o = \theta_o = 0$ and $u_{o-\Delta t} = u_{o+\Delta t}$ and $\theta_{o-\Delta t} = \theta_{o+\Delta t}$, where $0+\Delta t$ means next and $0-\Delta t$ previous time step. The Δt considered is 0.0001 s. For larger time steps, an unconditionally stable procedure, such as Newmark's method (e.g. Wood, 1990), need to be used. The results of calculation in Section 8 of Appendix are shown in Fig. 8.5. The calculated maximum block horizontal velocity is 1.7 mm/s, which is smaller than 3.3 mm/s that is calculated in Section 8 of Appendix using closed form solution – Equation (8.2) but without consideration of the block rotation θ. The calculated maximum horizontal velocity is also smaller than 6 mm/s specified by the manufacturer of the compressor for an accidental load on the foundation.

8.2.2 Energy Dissipation by Dampers

These devices are frequently used in combination with base isolators or on their own for retrofitting of vibration sources, which are not provided with base isolators. Soong and Dargush (1997), for example list the following types of passive dampers:

- Metallic
- Friction

- Viscoelastic
- Viscous fluid
- Tuned mass
- Tuned liquid

Passive dampers transform mechanical vibration energy into other forms of energy, mostly heat caused by friction within these dampers.

8.2.2.1 Example of Viscoelastic Dampers Effect on the Motion of a Foundation

It is interesting to see the effect of addition of viscous dampers in parallel with the rubber bearings considered in the case study in Section 8.2.1.1. Equation (8.1) of the horizontal block motion becomes:

$$m \cdot \frac{d^2u}{dt^2} + c \cdot \frac{du}{dt} + K_h \cdot \left(u - \theta \cdot \frac{3.84}{2}\right) = F \cdot \sin(2\pi \cdot f \cdot t) \ i.e.$$

$$\frac{d^2u}{dt^2} + 2 \cdot \xi \cdot \omega \cdot \frac{du}{dt} + \omega^2 \cdot \left(u - \theta \cdot \frac{3.84}{2}\right) = F \cdot \frac{\sin(2\pi \cdot f \cdot t)}{m} ,$$

$$(8.7)$$

where ξ is damping ratio $= c \, \omega \, (2 \, K_h)^{-1}$, c is damping coefficient and the other symbols are described for Equation (8.1). The closed form solution of the non-homogeneous second order differential equation (8.7) with constant coefficients is for the under damped case (when damping is less than the critical damping) and without consideration of the block rotation θ (e.g. Kramer, 1996):

$$u = e^{-\xi \cdot \omega \cdot t} \cdot [C_1 \cdot \sin(\omega_d \cdot t) + C_2 \cdot \cos(\omega_d \cdot t)] + \frac{F}{K_h} \cdot \frac{1}{(1 - \beta^2)^2 + (2 \cdot \xi \cdot \beta)^2}$$

$$\cdot [(1 - \beta^2) \sin(2\pi \cdot f \cdot t) - 2 \cdot \xi \cdot \beta \cdot \cos(2\pi \cdot f \cdot t)],$$

$$(8.8)$$

where $\omega_d = \omega \, (1 - \xi^2)^{1/2}$, β is defined for Equation (8.2). At $t = 0$, $u = 0$,

$$C_2 = \frac{F}{K_h} \cdot \frac{2 \cdot \xi \cdot \beta}{(1 - \beta^2)^2 + (2 \cdot \xi \cdot \beta)^2}. \quad \text{At} \quad t = 0, \ du(dt)^{-1} = 0,$$

$C_1 = \frac{F}{K_h} \cdot \frac{\beta}{\sqrt{1-\xi^2}} \cdot \frac{\beta^2 - 1 + 2\xi^2}{(1-\beta^2)^2 + (2\cdot\xi\cdot\beta)^2}$. For $\xi = 0.6$ and without consideration of the block rotation, the maximum velocity in the horizontal direction $= 2.0$ mm/s according to the calculation results in Section 9 of Appendix. The frequency of free vibration of the foundation block in the horizontal direction is only $(1 - 0.6^2)^{1/2} = 0.8$ times smaller than the frequency without damping. Equation (8.3) of the rotational block motion becomes:

$$I \cdot \frac{d^2\theta}{dt^2} + c \cdot \left(\frac{2.85^2}{2} + \left(\frac{3.84}{2} \right)^2 \right) \cdot \frac{d\theta}{dt} - c \cdot \frac{3.84}{2} \cdot \frac{du}{dt} + K_v \cdot \left(\frac{2.85}{2} \right)^2 \cdot \theta$$

$$+ K_h \cdot \frac{3.84}{2} \cdot \left(\frac{3.84}{2} \cdot \theta - u \right) = 2.95 \cdot F \cdot \sin(2\pi \cdot f \cdot t)$$

$$(8.9)$$

After substitution of u from Equation (8.8) and $du/(dt)^{-1}$ into Equation (8.9), it would be rather complicated to solve it in a closed form. Instead, a finite difference method (e.g. Wood, 1990) will be used with $\theta = d\theta/(dt)^{-1} = 0$ at $t = 0$ and $c = 2\xi$ $(K_h \, m)^{1/2}$. Equations (8.7) and (8.9) become:

$$m \cdot \frac{u_{next} - 2 \cdot u_{current} + u_{previous}}{(\Delta t)^2} + 2 \cdot \xi \cdot \sqrt{K_h \cdot m} \cdot \frac{u_{next} - u_{previous}}{2 \cdot \Delta t}$$

$$+ K_h \cdot \left(u_{current} - \theta_{current} \cdot \frac{3.84}{2} \right) = F \cdot \sin(2\pi \cdot f \cdot t_{current})$$

$$I \cdot \frac{\theta_{next} - 2 \cdot \theta_{current} + \theta_{previous}}{(\Delta t)^2} + 2 \cdot \xi \cdot \sqrt{K_h \cdot m} \cdot \left(\frac{2.85^2}{2} + \left(\frac{3.84}{2} \right)^2 \right)$$

$$(8.10)$$

$$\cdot \frac{\theta_{next} - \theta_{previous}}{2 \cdot \Delta t} - 2 \cdot \xi \cdot \sqrt{K_h \cdot m} \cdot \frac{3.84}{2} \cdot \frac{u_{next} - u_{previous}}{2 \cdot \Delta t}$$

$$+ \left(K_v \cdot \left(\frac{2.85}{2} \right)^2 + K_h \cdot \left(\frac{3.84}{2} \right)^2 \right) \cdot \theta_{current}$$

$$= 2.95 \cdot F \cdot \sin(2\pi \cdot f \cdot t_{current}) + K_h \cdot \frac{3.84}{2} \cdot u_{current}$$

For $\xi = 0.6$ and $\Delta t = 0.0001$ s, the results of calculations in Section 9 of Appendix are shown in Fig. 8.6. For larger time steps, an unconditionally stable procedure, such as Newmark's method (e.g. Wood, 1990), need to be used.

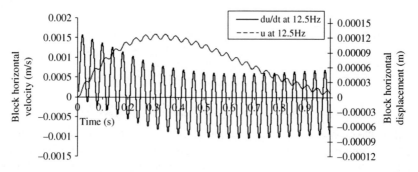

Fig. 8.6 Block horizontal velocity and displacement versus time in the case study in Section 8.2.2.1

The calculated maximum bock horizontal velocity is 1.5 mm/s, which is slightly smaller than 1.7 mm/s, which is calculated in Section 8.2.1.1 for the rubber bearing without viscous dampers, and therefore the use of viscous dampers is not needed in this case.

8.3   Ground Wave Propagation Barriers

When it is not possible to minimize vibration effects at the vibration source (due to restricted access for example), vibration propagation of Rayleigh, and other surface waves, towards recipients can be minimized by using barriers. The surface waves and their properties (displacement and velocity) are described in Section 2.3.2. For explanation how wave barriers work it is necessary to consider first wave properties such as the wave amplitude change with depth, Fig. 8.7.

A negative amplitude ratio indicates that the displacement at depth (z) is in the opposite direction of the surface displacement (u_o). The wave propagation velocity ranges from less than 100 m/s in very loose/soft soil to more than 700 m/s in very dense/stiff soil (e.g. Eurocode 8, Part 1, Table 3.1 for transversal wave velocity, which is similar to Rayleigh wave velocity). The frequency of Rayleigh wave could vary from about 1 Hz (e.g. Figs. 3.4 and 3.5) to several kHz in rock. Wave frequency can be measured as described in Section 3.2.1.1. In soil, wave length (as the ratio between the wave velocity and its frequency) could be a few tens of meters. The wave barriers are more effective when Rayleigh wave length is shorter and hence the wave amplitudes decrease more rapidly with depth.

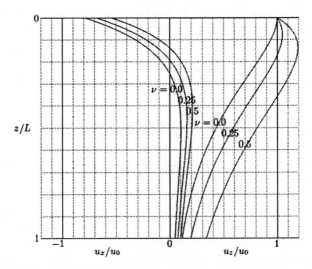

Fig. 8.7 Rayleigh wave amplitude ratio change in the horizontal (u_x/u_o) and the vertical direction (u_z/u_o) with respect to the amplitude at the surface (u_o) and the ratio z/L between depth (z) and the wave length (L) (Verruijt, 1994)

Fig. 8.8 Sketch of motion in the horizontal direction of a stiff barrier along its depth during passage of a Rayleigh wave

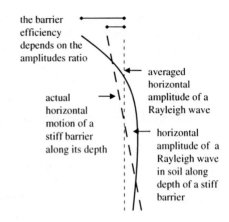

the barrier efficiency depends on the amplitudes ratio

actual horizontal motion of a stiff barrier along its depth

averaged horizontal amplitude of a Rayleigh wave

horizontal amplitude of a Rayleigh wave in soil along depth of a stiff barrier

8.3.1 Stiff Barriers

Stiff wave barriers act to average wave amplitudes over the barrier length as it has been described for kinematic soil-foundation interaction in Section 5.2. The averaged values are smaller than the peak values. Unfortunately, the horizontal amplitudes are not only averaged but also the wave amplitude modification result in rotation of a stiff barrier, which decreases the barrier's efficiency in the horizontal direction as sketched in Fig. 8.8. Prevention of variation of the wave amplitude along a stiff barrier depth causes occurrence of tensile stresses within the barrier.

8.3.1.1 Case Study on the Use of a Simplified Approach for Checking of the Effectiveness of a Pre-cast Concrete Wall Barrier

Nabeshima et al. (2004) presented the results of measurement, numerical modelling and field tests performed to assess effectiveness of a precast concrete wall barrier. The wall was made of 300 12 m long precast concrete hollow piles driven with the help of ground extraction by an auger. The top sandy soil to a depth of 10 m has a range of blow counts by standard penetration test between 2 and 20. Below the top soil layer, a 22.5 m thick soft clay layer has the blow count of SPT smaller than 5. The wall is located 8.5 m away from the vibration source, which is a 23 t truck that caused peak vertical acceleration of 4.5 cm/s^2 at the source with the predominant frequency of 10 Hz. This peak acceleration corresponds to the peak velocity of 4.5 $(2\pi 10)^{-1} = 0.07$ cm/s. For the Rayleigh wave velocity of about 150 m/s and the frequency of 10 Hz, the wave length is 150 $(10)^{-1} = 15$ m. For the barrier depth of 12 m, the depth to wave length ratio is 12 $(15)^{-1} = 0.8$, which means that the vertical amplitude of ground waves that passed the barrier decreased to $0.5 \times (100\% + 30\%) = 65\%$ of the vertical wave amplitude without the barrier according to Fig. 8.7 for Poisson's ratio of 0.25. The largest recorded accelerations were in the vertical direction and therefore the recorded and estimated vertical ground accelerations are shown in Fig. 8.9.

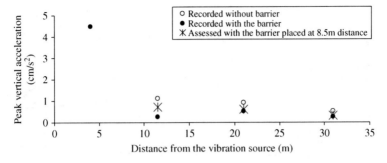

Fig. 8.9 Recorded and predicted peak acceleration versus distance for a stiff barrier in the case study in Section 8.3.1.1

8.3.2 Soft Barriers

Cut-of trenches without infill or with a soft infill physically prevent propagation of near surface (mainly Rayleigh) waves. However, the waves pass under the barriers and cause attenuated ground vibration behind such barriers. As the vertical amplitude of Rayleigh wave (Fig. 8.7) decreases approximately inversely proportionally to the ratio between depth and wave length and the wave amplitude decreases inversely proportionally to the horizontal distance from a vibration source (Equations 2.15 and 2.18) it follows that soft barriers effectively extent the horizontal source to receiver distance. The extended horizontal distance is a function of trench depth.

8.3.2.1 Case Study on the Use of a Simplified Approach for Checking of the Effectiveness of a Cut-Off Trench

Richart et al. (1970) described the results of field tests performed to evaluate the effectiveness of cut-off trenches of different depths and lengths, which are placed at 1.5 m distance from a vibration source with frequency range from 200 to 350 Hz. The trenches were located in a layer of uniform silty fine sand with the longitudinal wave velocity of 287 m/s. The underlying layer contains sandy silt with the longitudinal wave velocity of 534 m/s. For the frequency range from 200 to 350 Hz and Rayleigh wave velocity range from 137 to 117 m/s, Rayleigh wave length range is from 0.69 to 0.34 m. For the top layer unit density of 1665 kg/m^3, the maximum shear modulus range is from 31.2 to 22.8 MPa. Using Equation (2.15) and the equivalent distances equal to the actual horizontal distances plus twice the trench depth of 0.82 m for the places beyond the trench location, the calculated and recorded amplitudes of vertical displacements are shown in Fig. 8.10 for the trench depth to the wave length ratio of $1.19 = 0.82 (0.69)^{-1}$ and shear modulus of 31.2 MPa.

In this case, using actual distance plus twice the trench depth (as the waves passed around the trench) provided a good match with the recorded data at the distances greater than 1.5 m.

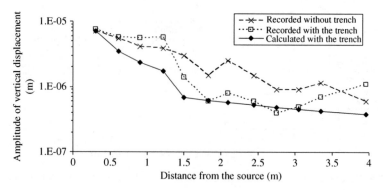

Fig. 8.10 Amplitudes of vertical ground displacements caused by the vibration source in the case study in Section 8.3.2.1

8.4 Recipient Isolators and Energy Dampers

When it is not possible or desirable to minimize propagation of ground vibration along the wave path, the last remedial measure is base isolation and/or energy damping of incoming ground vibration at the location of a recipient. Base isolation and/or energy damping of vibration sources and recipients is similar. The only difference is that in the former case waves are outgoing from a source and in the later case waves are incoming towards a recipient.

8.4.1 Passive Systems

Types of base isolators are listed in Section 8.2.1 and of energy dampers in Section 8.2.2.

8.4.1.1 Case Study of Isolation of a Building in Japan by Rubber Bearings

Hamaguchi et al. (2006) described seismic isolation of a nine story building above ground floor by twelve rubber bearings, which are located at the top of the ground floor columns. The building is 16.5 m wide, 25.7 m long and 30.3 m high above ground level and 2.3 m below the level. The superstructure is reinforced concrete rigid frame. The mass of storeys above the bearings is 3894 t, which is located 13.7 m above the bearings. The bearings are located under the columns, i.e. four along the length of 20.9 m and three along the width of 12.0 m. The properties of isolators are given in Table 8.2.

The shear stiffness and shear damping ratio change with shear strain within the isolation interface is shown in Fig. 8.11 (from Hamaguchi et al., 2006).

The shear stiffness K_h and damping ratio ξ dependence on shear strain γ of the isolation interface is:

Table 8.2 Properties of the isolators in the example in Section 8.4.4.1

Property	2 rubber bearings	4 lead rubber bearings	6 lead rubber bearings
Diameter (mm)	700	800	850
Height (mm)	287	373	368
Vertical stiffness (kN/m)	43340	26400	37920
Equivalent horizontal stiffness (kN/m)	–	2130	2400
Equivalent horizontal damping ratio	–	0.266	0.266

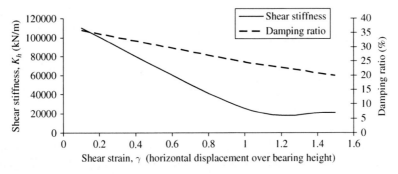

Fig. 8.11 Shear stiffness and damping ratio of the isolation interface in the example in Section 8.4.4.1

$$K_h = 20000 + \left(1 - \frac{\gamma}{1.1}\right) \cdot 100000, \quad \gamma \leq 1.1$$

$$K_h = 20000, \quad \gamma > 1.1 \qquad (8.11)$$

$$\xi = 0.2 + \left(1 - \frac{\gamma}{1.5}\right) \cdot 0.15, \quad \gamma \leq 3.5$$

Hamaguchi et al. (2006) modelled the structure as nine stacked lumped masses connected by shear springs with elastic-plastic three-linear hysteresis, while the base isolators are modelled as horizontal and rocking springs. The model of the building in this example is shown in Fig. 8.12

The equation of horizontal motion of the building model is:

$$m \cdot \frac{d^2u}{dt^2} + 2 \cdot \xi \cdot \sqrt{K_h \cdot m} \cdot \left(\frac{du}{dt} - \frac{d\theta}{dt} \cdot 13.7\right) + K_h \cdot (u - \theta \cdot 13.7)$$

$$= 2 \cdot \xi \cdot \sqrt{K_h \cdot m} \cdot \frac{du_g}{dt} + K_h \cdot u_g \qquad (8.12)$$

where $m = 2894$ t, K_h and ξ are according to Equation (8.11) in which $\gamma = |u - u_g|$ $(0.37)^{-1}$. Therefore, Equation (8.12) is non-homogeneous, non-linear second order differential equation. The equation of rotational motion of the building model is:

Fig. 8.12 Sketch of the building model in the example in Section 8.4.4.1

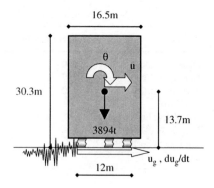

$$I \cdot \frac{d^2\theta}{dt^2} + 2 \cdot \xi \cdot \sqrt{K_h \cdot m} \cdot \left(\frac{12^2}{2} + 13.7^2\right) \cdot \frac{d\theta}{dt}$$

$$- 2 \cdot \xi \cdot \sqrt{K_h \cdot m} \cdot 13.7 \cdot \frac{du}{dt} + K_v \cdot \frac{12^2}{2} \cdot \theta + K_h \cdot 13.7 \cdot (13.7 \cdot \theta - u) \qquad (8.13)$$

$$= -2 \cdot \xi \cdot \sqrt{K_h \cdot m} \cdot 13.7 \cdot \frac{du_g}{dt} - K_h \cdot 13.7 \cdot u_g,$$

where $I = 12^{-1} \times 2894000 \times (30.3^2 + 16.5^2) = 287072742$ kg m^2, $m = 2894$ t, K_h and ξ are according to Equation (8.11) in which $\gamma = |u - u_g| \, (0.37)^{-1}$. Therefore, Equation (8.13) is non-homogeneous, non-linear second order differential equation. The results of the calculations may be chaotic. To simplify the calculations, Equations (8.12) and (8.13) will be solved for the upper and lower bound values of K_h and ξ only. Using an explicit algorithm, with very small time step (e.g. 0.0001 s in Section 8.2.1.1 and 8.2.2.1) would not be practical for rather long input motion lasting several tens of seconds. Instead, Newmark algorithm, which is unconditionally stable for any time step (e.g. Zienkiewich and Taylor, 1999), is used in the example. According to the algorithm:

$$u_{next} = u_{current} + \Delta t \cdot \frac{du_{current}}{dt} + \frac{\Delta t^2}{4} \cdot \frac{d^2 u_{current}}{dt^2} + \frac{\Delta t^2}{4} \cdot \frac{d^2 u_{next}}{dt^2}$$

$$\frac{du_{next}}{dt} = \frac{du_{current}}{dt} + 0.5 \cdot \Delta t \cdot \frac{d^2 u_{current}}{dt^2} + 0.5 \cdot \Delta t \cdot \frac{d^2 u_{next}}{dt^2}$$

$$m \cdot \frac{d^2 u_{next}}{dt^2} + 2 \cdot \xi \cdot \sqrt{K_k \cdot m} \cdot \left(\frac{du_{next}}{dt} - \frac{d\theta_{next}}{dt} \cdot 13.7\right)$$

$$+ K_h \cdot (u_{next} - \theta_{next} \cdot 13.7) = 2 \cdot \xi \cdot \sqrt{K_h \cdot m} \cdot \frac{du_{g,next}}{dt} + K_h \cdot u_{g,next}$$

$$\theta_{next} = \theta_{current} + \Delta t \cdot \frac{d\theta_{current}}{dt} + \frac{\Delta t^2}{4} \cdot \frac{d^2\theta_{current}}{dt^2} + \frac{\Delta t^2}{4} \cdot \frac{d^2\theta_{next}}{dt^2}$$

$$\frac{d\theta_{next}}{dt} = \frac{d\theta_{current}}{dt} + 0.5 \cdot \Delta t \cdot \frac{d^2\theta_{current}}{dt^2} + 0.5 \cdot \Delta t \cdot \frac{d^2\theta_{next}}{dt^2}$$

$$I \cdot \frac{d^2\theta_{next}}{dt^2} + 2 \cdot \xi \cdot \sqrt{K_h \cdot m} \cdot \left(\frac{12.2^2}{2} + 13.7^2\right) \cdot \frac{d\theta_{next}}{dt} - 2 \cdot \xi \cdot \sqrt{K_h \cdot m} \cdot 13.7$$

$$\cdot \frac{du_{next}}{dt} + K_v \cdot \frac{12.2^2}{2} \cdot \theta_{next} + K_h \cdot 13.7 \cdot (13.7 \cdot \theta_{next} - u_{next})$$

$$= -2 \cdot \xi \cdot \sqrt{K_h \cdot m} \cdot 13.7 \cdot \frac{du_{g,next}}{dt} - 13.7 \cdot K_h \cdot u_{g,next}$$

$$(8.14)$$

Without coupling of u and θ, the equations for u_{next} and $du_{next}\,(dt)^{-1}$ replaced into the third equation would provide solution for $d^2u_{next}(dt^2)^{-1}$, which would be used to calculate u_{next} and $du_{next}\,(dt)^{-1}$ from the first and second equation. Similarly, without coupling of u and θ, the equations for θ_{next} and $d\theta_{next}\,(dt)^{-1}$ replaced into the sixth equation would provide solution for $d^2\theta_{next}(dt^2)^{-1}$, which would be used to calculate θ_{next} and $d\theta_{next}\,(dt)^{-1}$ from the fourth and fifth equation. Because of the coupling between u and θ, it is necessary to use a predictive-corrective method to calculate the variables u_{next}, $du_{next}\,(dt)^{-1}$, θ_{next} and $d\theta_{next}\,(dt)^{-1}$. The predictive rotational acceleration is assumed to be the same as the previous rotational acceleration and the calculation repeated until the difference between the assumed and predicted value is less than 5% or the number of trials exceeds 10. The limits are set arbitrary. The calculation is given in Sections 10 and 11 in Appendix. The ground motion time histories used by Hamaguchi et al. (2006) were not available. Instead, similar time history obtained from Ambraseys et al. (2004) is used. The basic comparative properties of the time histories used are given in Table 8.3.

The results of the calculations are shown in Figs. 8.13 and 8.14.

The ratio between maximum calculated building horizontal acceleration and the input acceleration is 0.53 for the upper bound horizontal stiffness and damping ratio and 0.66 for the lower bound horizontal stiffness and damping ratio. Hamaguchi et al. (2006) reported for their model, the minimum ratio between the maximum horizontal acceleration of the top storey of the building and the maximum input acceleration of 0.64 and the maximum ratio between the maximum horizontal

Table 8.3 Basic comparative data of the replacement time history used in the example in Section 8.4.4.1

Description	Great Kanto fault model (Hamaguchi et al., 2006)	Replacement Yarimka-Petkim station NS record during the 17 August 1999 Izmit earthquake (Ambraseys et al., 2004)
Peak acceleration (m/s^2)	3.09	3.05
Peak velocity (cm/s)	49.9	55.9
Peak displacement (cm)	36.1	26.9

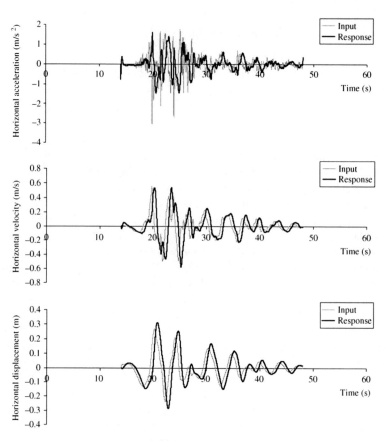

Fig. 8.13 Input base and calculated building time histories for the $K_h = 120000$ kN/m and $\xi = 0.35$ (the upper bound values) in the example in Section 8.4.4.1

acceleration of the top storey of the building and the maximum input acceleration of 0.88 (for the upper bound horizontal stiffness and damping). Zienkiewich and Taylor (1999), for example, provided details how to solve the problem of non-linear differential equations, which arises if the horizontal damping and damping ratio are considered to be dependent on the horizontal displacement.

8.4.2 Active Systems

Soong and Dargush (1997), for example, listed the following semi-active and active control systems:

- Active bracing systems
- Active mass dampers
- Variable stiffness or damping systems
- Smart materials

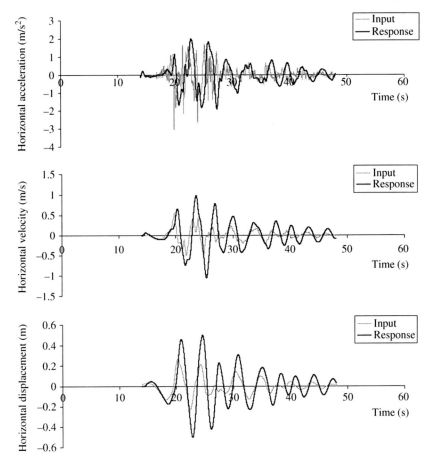

Fig. 8.14 Input base and calculated building time histories for the $K_h = 20000$ kN/m and $\xi = 0.20$ (the lower bound values) in the example in Section 8.4.4.1

The above systems are incorporated into structures, which vibration is outside the scope of this volume.

8.5 Summary

This chapter provides examples of the following vibration control measures:

- Preferable minimization of vibration at the source by base isolation and vibration energy dissipation by dampers.
- Provision of ground wave propagation barriers (stiff or soft – trenches) when the preferable measure is not possible.
- Use of passive systems for individual recipient isolation from incoming waves and for incoming energy damping.

Chapter 9
Effects on Soil Slopes and Shallow Foundations

9.1 Introduction

Ground vibration can change ground properties and cause ground slope instability, decrease of foundation bearing capacity or foundation failure if liquefaction occurs and increased settlement. Not many cases are reported concerning ground vibration effects on change of soil properties except for earthquakes, which cause greater amplitudes and smaller number of cycles than other sources of ground vibrations.

Scope of this chapter is to present simplified analyses of detrimental effect of ground vibration on slopes and shallow foundations. The effects of ground vibrations on retaining walls are considered for earthquakes by Srbulov (2008), for example, and can be used for the ground vibrations in general.

9.2 Slope Instability Caused by Vehicle Induced Ground Vibration

Slope stability in cyclic condition depends on soil shear strength, which depends on either effective axial stress (total stress less pore water pressure) and friction angle ϕ' of coarse grained soil or on cohesion c_u of fine grained soil (more than about 35% by weight of grain sizes less than about 0.05 mm diameter) in short term (undrained) conditions. The friction angles in cyclic condition during strong earthquakes and for soil zones in vicinity of pile driving, blasting by explosives and soil compaction are shown in Fig. 7.5 for sandy soil. Cohesion of normally consolidated clay in static condition is described by Equation (7.11), while the detrimental effect of a large number of cycles is shown in Fig. 7.8. Change of soil shear strength can be caused by cyclic loading of small amplitude but with large number of cycles such as described in Section 1.2.4.1 and 1.2.4.2.

When coarse grained soil is in loose to medium dense state it contains many voids between its grains. Vibration causes grains to move into existing voids in order to achieve the state of minimum energy. This causes decrease of soil volume and densification, which is beneficial during soil compaction by vibratory machines.

M. Srbulov, *Ground Vibration Engineering*, Geotechnical, Geological, and Earthquake Engineering 12, DOI 10.1007/978-90-481-9082-9_9, © Springer Science+Business Media B.V. 2010

Compacted soil is not saturated and compaction is achieved by decrease of air filled voids with existing fluid acting as a lubricant. When loose to medium dense coarse grained soil is saturated, its volume tends to decrease during vibration that in turn causes an increase in excess pore water pressure when soil drainage is prevented by impervious barriers or by long drainage paths in comparison with the rate of excess pore water pressure build-up. If built-up excess water pressure riches the value of acting total stress, the shear strength of soil decreases to near zero and such soil behave like fluid i.e. liquefies. If soil is partly saturated or medium dense to dense then it may not loose its shear strength completely but only partially, which is called residual shear strength.

Cohesion of fine grained soil is independent from axial total and effective stress in saturated condition and is probably caused by electrochemical forces acting within dense water films around individual particles (more details on the attractive and repulsive forces acting around fine grains are provided by Mitchell, 1993, for example). This is true because an application of direct electric current through such soil causes water drainage out of it and soil densification with increase in its cohesion (more details on this technique is provided by Van Impe, 1989, for example). On contrary, change of chemistry of water films around fine grains by removal of sodium cations causes breakage of existing chemical bonds and loss of cohesion i.e. piping type failures (more details on this process is provided by Bell and Culshaw, 1998, for example). Wave propagation through soft cohesive soil probably causes increased polarization i.e. further separation of electrons within water films around grains, which in turn results in smaller attraction forces and decrease of soil cohesion. Very sensitive clay in Norway and Canada can completely loose its cohesion and flow like a thick fluid when disturbed by ground vibration or due to overstressing (more information on sensitive clay is provided by Mitchell and Houston, 1969, for example).

In order to avoid problems caused by soil disturbance during its sampling and transportation to the laboratory and also by various effects of testing apparatus and its limitation in applying cyclic loading equivalent to actual loading in the field it is best to perform field tests using vehicles, which will cause cyclic loading of soil. Measurements of increase of excess pore water pressure within soil mass can be performed by piezo-cone (e.g. ASTM D5778-95, Eurocode 7 – Part 2, 2007) or vibrating wire piezometers (e.g. Hanna, 1985). Such field tests will provide functional relationship between excess pore water pressure built-up and number of cycles for a specific vehicle. Once both steady state and cyclic pore water pressures are defined it is possible to calculate factor of safety of slope stability F_s using, for example, the routine method by Bishop (1955) for circular cylindrical trial slip surfaces, Fig. 9.1.

$$F_s = \frac{\sum \left[c' \cdot b + (W_s - u_w \cdot b) \cdot \tan \phi' \right] \cdot \dfrac{1}{\cos \alpha_s + \sin \alpha_s \cdot \dfrac{\tan \phi'}{F_s}}}{\sum (W_s \cdot \sin \alpha_s)}, \qquad (9.1)$$

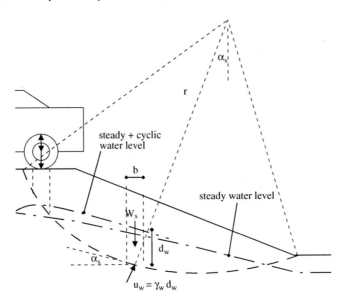

Fig. 9.1 Cross section through a circular cylindrical trial slip surface

where c' is cohesion in terms of effective stresses; static and dynamic load from a vehicle near slope crest is added to the weight W_s of corresponding slice with width b; u_w is steady and cyclic pore water pressure at the base of a slice, α_s inclination to the horizontal of a slice. Equation (9.1) contains F_s implicitly and is solved iteratively.

9.2.1 Case Study of the Instability of Asele Road Embankment in Sweden

The embankment failure is described in Section 1.2.4.1. The 7.5 m high embankment had the crest width of about 7.5 m and the slopes inclined at 1.5 horizontal to 1 vertical. The slopes were protected by 1 m thick layer of quarry rock placed over 0.5 m thick gravel layer. The steady water level was about 2 m below the crest level at the time of failure. The vibratory roller was located close to the slope crest. The standard penetration test blow count was about 7 within the embankment made of fine grained till. The remains of the embankment indicate deep failure surface or a retrogressive type failure. Search for minimum F_s in steady state condition is performed using computer software (e.g. Maksimovic, 1988) and Bishop (1955) method. For steady state $F_s = 1$, the embankment till is necessary to have $c' = 10$ kPa and $\phi' = 35°$. On the liquefaction of embankment fill, its frictional angle ϕ' would be only $5°$ according to Fig. 7.5, for the $(N_1)_{60} = 7$ and the fines content of 35%. An estimation of the parameters of slope movement is performed for its equivalent sliding block shown in Fig. 9.2.

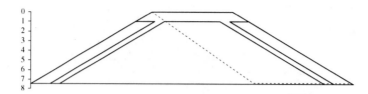

Fig. 9.2 A cross section through the Asele Road embankment with an equivalent sliding block (*dashed lines*)

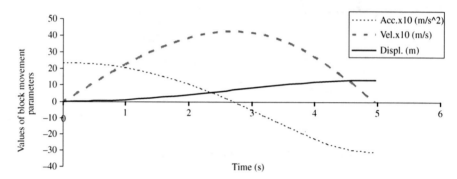

Fig. 9.3 Parameters of equivalent translational block movement for the case study in Section 9.2.1

The analysis of movement of the translational sliding block on two basal planes is performed according to Ambraseys and Srbulov (1995). Details of the calculations are given in Section 12 of Appendix. The block thickness is 4.1 m and its length 13.5 m based on the geometry shown in Fig. 9.2; the block angle of inclination to the horizontal is adopted 33.7° (for 1.5 horizontal to 1 vertical slope inclination). If the friction angle of embankment till had been decreased during cyclic loading to 10.5°, the calculation shows that the block would have completely slid into the lake within 5 s time interval with maximum velocity of 4.25 m/s, Fig. 9.3.

9.3 Shallow Foundation Settlement Caused by Ground Vibration

Loose to medium dense coarse grained soil decreases its volume (becomes compact) when undergoes vibration due to its particles movement into the voids surrounding soil grains. Degree of soil compaction can be described by its relative density D_r, which is calculated using the formula (e.g. Das, 1985):

$$D_r = \frac{\gamma_{d(max)}}{\gamma_d} \cdot \frac{\gamma_d - \gamma_{d(min)}}{\gamma_{d(max)} - \gamma_{d(min)}}, \tag{9.2}$$

where maximum, minimum and natural soil dry density γ_d of coarse grained soil can be determined using, for example, ASTM D4914, D1556, D4253, D4254,

Table 9.1 Relationships between N_{SPT} and relative density (D_r)

N_{SPT}	Density	D_r (%)
0–4	Very loose	< 15
4–10	Loose	15–35
10–30	Medium dense	35–65
30–50	Dense	65–85
> 50	Very dense	> 85

BS1377-2, BS1377-4. Relative density can also be estimated from the results of filed tests. For the standard penetration test blow count N_{SPT}, Terzaghi and Peck (1948), for example, suggested the classification shown in Table 9.1.

Lunne et al. (2001) provide graphs for estimation of relative density based on cone penetration tests. Relative density can be expressed as well in terms of void ratio e, i.e. the ratio between volume of voids and volume of solids within soil.

$$D_r = \frac{e_{max} - e}{e_{max} - e_{min}},$$

$$e_{min} = \frac{G_s \cdot \gamma_w}{\gamma_{d(max)}} - 1,$$

$$e_{max} = \frac{G_s \cdot \gamma_w}{\gamma_{d(min)}} - 1 \qquad (9.3)$$

where G_s is the specific gravity of soil solids ($\sim$2.65) and γ_w is the unit weight of water 9.81 kN/m³. Void ratio e is related to soil porosity n as $e = n\,(1-n)^{-1}$, or in turn $n = e\,(1+e)^{-1}$. Soil porosity n is defined as the ratio of the volume of voids (filled with air and water) to the total volume of soil. Decrease in soil porosity Δn is caused by decrease of voids in soil and is equal to the volumetric deformation ε_v of soil. For a unit area of soil, ε_v equals to the vertical strain ε, which when integrated along depth of interest equals to soil settlement. Ground vibration causes transmission of wave energy through soil under a foundation. The increase in energy ΔE of soil at a depth beneath foundation can be calculated as a product of acting total axial vertical stress times the active area under a foundation times the wave amplitude times the number of cycles. From a graph between soil porosity and the energy applied to achieve such porosity and calculated ΔE, Fig. 9.4, it is possible to define $\Delta n = \varepsilon$ and from it to calculate soil settlement caused by ground vibration as an integral of ε over depth of interest.

Acting axial vertical stress at a depth is a sum of the total overburden pressure caused by soil weight above that depth and the additional total vertical stress due to foundation load. The additional vertical stress $\Delta\sigma_{v,z}$ due to foundation load at a depth z can be calculated either using Boussinesq (1885) formula or from a truncated cone. According to Boussinesq (1885):

$$\Delta\sigma_{v,z} = \frac{3 \cdot V \cdot z^3}{2 \cdot r^5},$$

$$r = \sqrt{z^2 + d_h^2} \qquad (9.4)$$

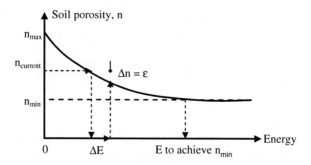

Fig. 9.4 Soil porosity versus energy

Fig. 9.5 Truncated cone
model

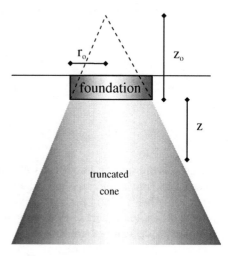

where V is vertical force acting at the foundation underside, d_h is the horizontal
distance between the location where the load V is acting and the location where the
stress is calculated. A truncated cone (e.g. Wolf, 1994) is shown in Fig. 9.5.

According to Wolf (1994), for example,

$$\frac{z_o}{r_o} = \frac{\pi}{4} \cdot (1 - v) \cdot \left(\frac{v_c}{v_t}\right)^2, \tag{9.5}$$

where v is Poisson's ratio, v_c is wave velocity, which is equal to the velocity of
longitudinal waves v_l when $v \leq 1/3$ and $v_c = 2v_t$ when $1/3 < v < \frac{1}{2}$, where v_t is
transversal wave velocity. From the geometry of Fig. 9.5 it follows that

$$r = r_o \cdot \left(1 + \frac{z}{z_o}\right)$$

$$\Delta\sigma_{v,z} = \Delta\sigma_{v,z_o} \cdot \left(\frac{r_o}{r}\right)^2 \tag{9.6}$$

where r_o is the radius of an equivalent disk, $r_o = (A_f \pi^{-1})^{1/2}$, A_f is actual foundation area. Ground wave amplitude at the ground surface is calculated from Equation (2.15) or (2.16) and estimated at a depth z from Fig. 8.7 for Rayleigh waves in the free field. For body waves, Equations (2.11), (2.12), and (2.13) will provide wave velocity amplitude, from which the wave displacement amplitude can be calculated by dividing the velocity by $2\pi f$, where f is the frequency of vibration. The number of wave cycles depends on the number of cycles of a vibration source.

9.3.1 Case Study of Foley Square Building Settlement Caused by Pile Driving in Its Vicinity

Lacy and Gould (1985) described several cases of the settlements of buildings on shallow and piled foundations, of ground surface and sewer pipes as a result of pile driving in narrowly graded, single sized clean sand with relative density less than about 50 to 55% even if the recorded peak particle velocities were much smaller than 50 mm/s, which is considered to be a safe limit for buildings. Lacy and Gould (1985) conclude that factors (the number, length and type of piles and driving resistance) that increase the total vibration energy input will increase settlements.

Foley Square building in Manhattan – New York City experienced settlement due to 14HP73 piles driving through 24 m of sand and silt. The building is 49.5 m long and of unspecified width. First, a 2.5 cm settlement of the building occurred as a result of the use of impact hammer with the input energy of 35.2 kJ per blow, with 22–40 blows per 0.305 m penetration causing the peak particle velocity of 4.8 mm/s at a distance of 6.1 m from the pile. Consequently used 'subsonic' pile hammer with 29 blows per 0.305 m penetration caused the peak particle velocity of 3.5 mm/s at a distance of 6.1 m from the pile and a Bodine'-'sonic' pile hammer caused the same peak particle velocity at the same distance as the impact pile. Final measured settlements of the foundations reached 7.6 cm. Ground between the piles and the building settled 0.3 m. The bearing pressure under the 16 story building was quoted to be 480 kPa by Lacy and Gould (1985). However, it may be under individual footings as an expected overall pressure should be about $16 \times 5 = 90$ kPa. The seismographs indicated a typical frequency of the glacial fine sand of 30 Hz. The recorded standard penetration blow count in sand and silt varied in the range from 22 to 50, with typical 29 blows under the building to a depth of 30.5 m. Lacy and Gould (1985) estimated initial relative density of sand and silt at the site of about 45%.

The graph in Fig. 9.4 is not available for this case. Based on data by Das (1985), it is assumed that for silty sand $e_{max} = 1$ ($n_{max} = 0.5$) and $e_{min} = 0.4$ ($n_{min} = 0.28$). For Δn_{max} of $0.5 - 0.28 = 0.22 = \varepsilon_{max}$, the soil compression over 1 m depth d is 0.22 m and the energy necessary to achieve such settlement over a unit area A of 1 m^2 is force × displacement $= \varepsilon_{max} * E' * A * \varepsilon_{max} * d = 0.22 * 29 * 1 * 0.22 * 1 = 1.40$ MJ, where the soil modulus $E' = N_{SPT}$ in MPa is assumed based on data by Stroud (1988), for example. For assumed a simple circular function with the horizontal tangent at the point ($E = 1.40$ MJ, $n = 0.28$), it follows that

$$n = 4454545.85 - \sqrt{4454545.565^2 - (E - 1400)^2}$$
$$E = 1400 - \sqrt{4454545.565^2 - (n - 4454545.845)^2},$$
(9.7)

where energy E is in kJ. For estimated initial $D_r = 0.45$, $e = 0.73$, $n_{current} = 0.42$, $E_{current} = 283.2$ kJ. For assumed the building width equal to the building length $r_o = 27.9$ m. For $N_{SPT} \sim 29$ and Equation (6.1), $v_t = 246$ m/s. For assumed $v = 0.3$ and Equation (2.5), $v_l = 460$ m/s. From Equation (9.5), $z_o = 53.6$ m. The recorded peak particle velocity of 0.0048 m/s at the ground surface at the frequency of 30 Hz corresponds to the amplitude of ground displacement of 0.0048 $(2\pi\ 30)^{-1} = 0.0000255$ m. The Rayleigh wave length was $\lambda = 246\ (2\pi\ 30)^{-1} = 1.3$ m. The number of cycles over 24 m depth was $24*(0.305)^{-1}*30$ blows/feet = 2360 cycles for an averaged recorded 30 blow/feet of the driving hammers. The results of the calculations of settlement without the soil-structure interaction effects considered are given in Table 9.2. The wave amplitudes along depths are obtained from Fig. 8.7.

The calculated settlement of 0.19 m is greater than the recorded settlement of the building of 0.025 m after driving of the first pile possibly due to the assumptions made in the calculation and/or the soil-structure interaction effect and/or considering the change of wave amplitudes with depth according to Fig. 8.7, which is applicable to the free field without structures. The effect of kinematic soil-structure interaction that is described in Section 5.2 has not been considered in this case study because only amplitude and not time history of ground motion is available. If no stress from the building is considered (i.e. $\Delta\sigma_v = 0$ in the free field), the calculated settlement is 0.03 m (Table 9.3), which is quite close to the recorded settlement of 0.025 m after driving of the first pile.

9.4 Bearing Capacity of Shallow Foundation over Liquefied Soil Layer

Determination of liquefaction potential of soil under foundation due to vibration is best performed using cyclic simple shear tests described in Section 6.4.2. The test conditions used should be as close as possible to the field conditions. As shallow foundation is usually placed above ground water level, soil liquefaction could occur at some distance from foundation underside. In such a case, punch through type failure may occur. If punch through type failure occurs, a shallow foundation can sink until the buoyancy force acting on the punched through wedge in Fig. 9.6 comes into equilibrium with the applied load. The vertical foundation capacity F_v in the case of punch through failure is (e.g. SNAME, 1997)

$$F_v = F_{v,b} - A_f \cdot H_l \cdot \gamma + 2 \cdot \frac{H_l}{B_f}(H_l \cdot \gamma + 2 \cdot p'_o)K_s \cdot \tan\phi \cdot A_f, \qquad (9.8)$$

where $F_{v,b}$ is determined assuming the foundation bears on the surface of the lower liquefied layer, A_f is foundation area, H_l is distance from foundation level

Table 9.2 Results of the calculation of the settlement under the building due to pile driving in the case study in Section 9.3.1

z (m)	r (m)	$r^2\pi$ (m²)	σ_v (kPa)	$\Delta\sigma_v$ (kPa)	$\sigma_v+\Delta\sigma_v$ (kPa)	$V=(\sigma_v+\Delta\sigma_v)*r^2\pi$ (kN)	z/λ	$\Delta_{w,z}$ (m)	$\Delta E = V*\Delta_{w,z}*2360$ (kJ)	$E_{current}+\Delta E$ (kJ)	n	$\varepsilon=\Delta n$
1	28.4	634	10	87	96	61049	0.8	7.65E-06	1102	1385	0.29	0.13
2	28.9	658	29	84	112	73771	1.5	1.28E-06	222	505	0.37	0.05
3	29.5	682	48	81	128	87404	2.3	2.55E-07	53	336	0.41	0.01
4	30	706	67	78	144	101973	3.1	1.28E-07	31	314	0.42	0.00
5	30.5	731	86	75	161	117501	3.8	2.60E-08	7	290	0.42	0.00
6	31	756	105	73	177	134014	4.6	1.30E-08	4	287	0.42	0.00
7	31.5	781	124	70	194	151534	5.4	–	–	–	–	–

Settlement (m) = 0.19

Table 9.3 Results of the calculation of the settlement in the free field due to pile driving in the case study in Section 9.3.1

z (m)	r (m)	$r^2\pi$ (m²)	σ_v (kPa)	$\Delta\sigma_v$ (kPa)	$\sigma_v+\Delta\sigma_v$ (kPa)	$V=(\sigma_v+\Delta\sigma_v)*r^2\pi$ (kN)	z/λ	$\Delta_{w,z}$ (m)	$\Delta E = V*\Delta_{w,z}*2360$ (kJ)	$E_{current}+\Delta E$ (kJ)	n	$\varepsilon=\Delta n$
1	28.4	634	10	0	10	6027	0.8	7.65E-06	109	392	0.40	0.02
2	28.9	658	29	0	29	18748	1.5	1.28E-06	56	340	0.41	0.01
3	29.5	682	48	0	48	32381	2.3	2.55E-07	19	303	0.42	0.00
4	30	706	67	0	67	46950	3.1	1.28E-07	14	297	0.42	0.00
5	30.5	731	86	0	86	62478	3.8	2.60E-08	4	287	0.42	0.00
6	31	756	105	0	105	78991	4.6	1.30E-08	2	286	0.42	0.00
7	31.5	781	124	0	124	96512	5.4	–	–	–	–	–

Settlement (m) = 0.03

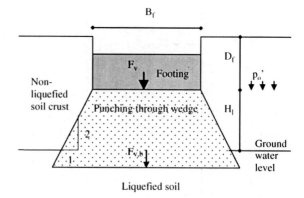

Fig. 9.6 Punching through mode of failure due to sub-layer liquefaction

to the level of liquefied layer below, γ is unit weight of non-liquefied soil, p_o' is effective overburden stress at the foundation depth, K_s is the coefficient of punching shear that is calculated from the equation $K_s \cdot \tan\phi = \frac{3 \cdot c_u}{B_f \cdot \gamma}$, ϕ is friction angle of non-liquefied layer, c_u is undrained shear strength of liquefied sand layer (= 0.05 to 0.12 or 0.09 on average of the effective overburden pressure according to Olson and Stark, 2002), B_f is diameter of an equivalent circular foundation and $F_{v,b} = (c_u\,N_c + p')\,A_f$, where N_c =5.14, p' effective overburden stress at the top of liquefied layer.

9.5 Summary

This chapter contains simplified analyses with case studies of ground vibration effects on soil properties resulting in:

- Slope instability
- Shallow foundation settlement
- Punching through type failure

Appendices – Microsoft Excel Workbooks on http://extras.springer.com

The MS Excel spreadsheet format is used for maximum portability. Microsoft provides MS Excel Viewer free of charge at its Internet web site.

The spreadsheets are kept as simple as possible.

If MS Excel complains at the start about the security level of macros please click on Tools then Macro then Security and adjust the security level to at least medium. The spreadsheet must be exited and re-entered for the change made to take place.

The spreadsheets are applicable to the vase studies and examples considered in this monograph.

1 Fast Fourier Transform, Filtering and Inverse Fast Fourier Transform

The work book is referred in Section 4.2.1.1. A work sheet from the work book is shown in Fig. 1

2 Polynomial Base Line Correction

The work book is referred in Section 4.3.1. A work sheet from the work book is shown in Fig. 2

3 Elastic Response Spectra of a Single Degree of Freedom Oscillator

The work book is referred in Section 4.4.3.1. A work sheet from the work book is shown in Fig. 3

4 Peak Particle Velocities from Piles Driving

The work book is referred in Section 7.2.1.2. A work sheet from the work book is shown in Fig. 4

M. Srbulov, *Ground Vibration Engineering*, Geotechnical, Geological, and Earthquake Engineering 12, DOI 10.1007/978-90-481-9082-9, © Springer Science+Business Media B.V. 2010

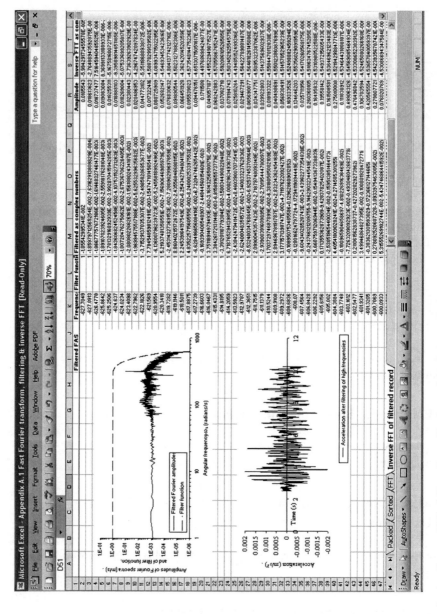

Fig. 1 Works sheet Inverse FFT of filtered record in the work book Appendix 1

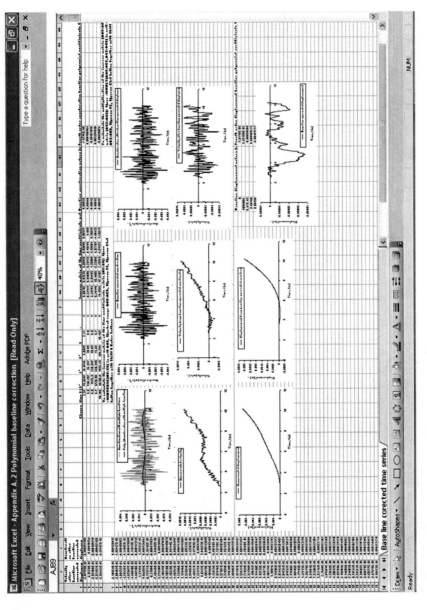

Fig. 2 Work sheet Base line corrected time series in the work book Appendix 2

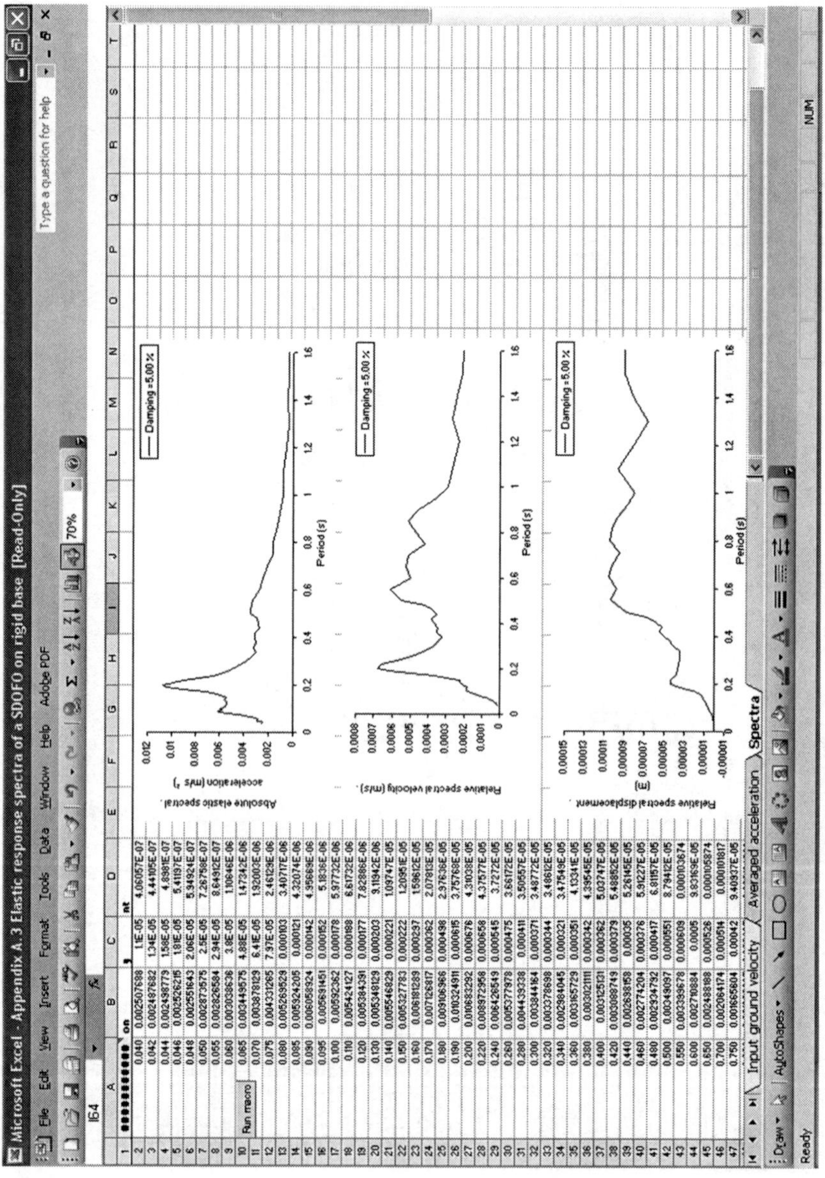

Fig. 3 Work sheet Spectra in the work book Appendix 3

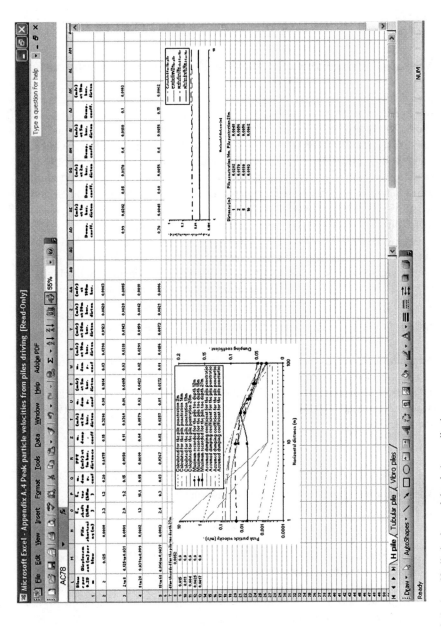

Fig. 4 Work sheet H pile in the work book Appendix 4

5 Peak Particle Velocities from Vibratory Rollers

The work book is referred in Section 7.2.2.2. A work sheet from the work book is shown in Fig. 5

6 Vibration Properties of a Shallow Foundation for Compressor

The work book is referred in Section 7.4.2. A work sheet from the work book is shown in Fig. 6

Wolf (1994) described a discrete element model for coupled rocking and horizontal displacement of foundation of a three-cylinder compressor. The 2D discrete element model is shown in Fig. 7.52. Soil reaction to foundation movement is considered in the horizontal direction and in rotation by elastic springs and dashpots. The elastic spring and dashpot with negative coefficients are artificial and are introduced by Wolf (1994). The two triangles under the foundation represent trapped soil beneath foundation for Poisson's ratio greater than 1/3 (Wolf, 1994).

The relationship for soil reaction moment contains a convolution integral in time. As an alternative to the recursive evaluation of the convolution integral, a physical discrete element model, which incorporates rigorously the convolution implicitly, is used according to Wolf (1994). The equation of the model rotational motion is:

$$(\Delta M_\theta + I) \cdot \ddot{\theta}_o + K_\theta \cdot \theta_o - \frac{K_\theta}{3} \cdot (\theta_o - \theta_1) + C_\theta \cdot \dot{\theta}_o - C \cdot e \cdot \dot{u}_o - K \cdot e \cdot u_o = M \quad (1)$$

An additional internal rotational degree of freedom located within the foundation soil and connected by a rotational spring with a coefficient to the base and by a rotational dashpot with a coefficient to the rigid support, is introduced by Wolf (1994). Both the negative coefficients are artificial and are introduced by Wolf (1994) for mathematical reason.

$$-\frac{K_\theta}{3} \cdot (\theta_1 - \theta_o) - C_\theta \cdot \dot{\theta}_1 = 0 \quad (2)$$

The equation of the model translational motion is:

$$m \cdot (\ddot{u}_o + e \cdot \ddot{\theta}_o) + C \cdot \dot{u}_o + K \cdot u_o = 0 \quad (3)$$

The symbols used in Equations (1), (2), and (3) are: ΔM_θ is the trapped soil mass moment of inertia when soil Poisson's ratio is greater than 1/3 (Equation 4), I is the foundation mass moment of inertia around the centre of gravity $= (12)^{-1} \times$ footing mass $\times$ (footing width^2 + footing height^2), θ_o is the angle of footing rotation, K_θ is the rotational static stiffness coefficient (Equation 5), θ_1 is an additional internal rotational degree of freedom, C_θ is the rotational dashpot coefficient (Equation 5), C is the translational dashpot coefficient (Equation 6), e is the distance between footing centre of gravity and its base, u_o is the horizontal footing

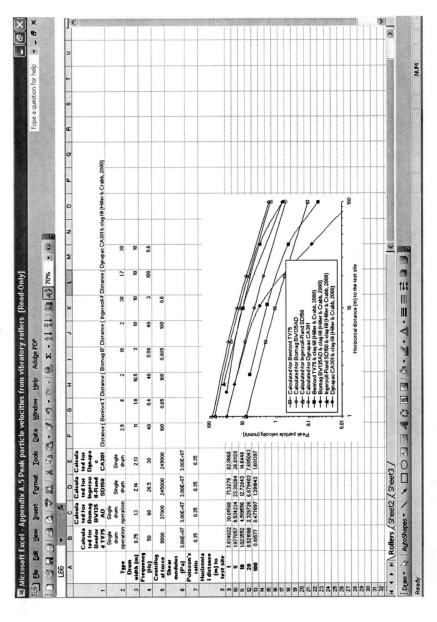

Fig. 5   Work sheet Rollers in the work book Appendix 5

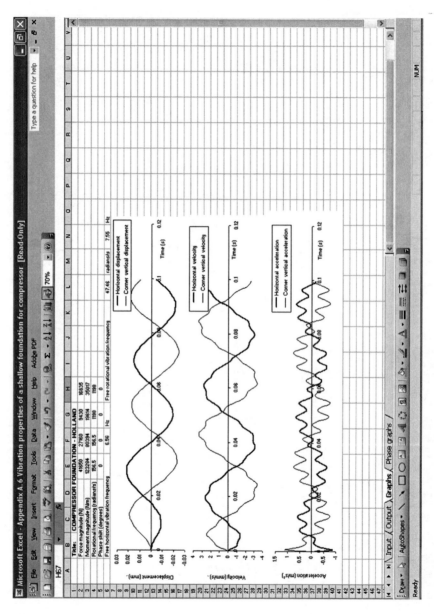

Fig. 6 Work sheet Graphs in the work book Appendix 6

displacement, K is the translational static stiffness coefficient (Equation 6), M is the rotational moment around the footing centre of gravity, m is the footing mass,; dot and double dot above a variable represent the first and second derivative in time.

$$\Delta M_\theta = 0.3 \cdot \pi \cdot \left(v - \frac{1}{3}\right) \cdot \rho \cdot \left(\frac{l_c \cdot b_c^3}{3 \cdot \pi}\right)^{\frac{5}{4}} \tag{4}$$

For assumed surface foundation on homogeneous half space and footing width b_c and length l_c, the coefficients according to Wolf (1994) for footing rotation along (around) l_c are:

$$K_\theta = \frac{G \cdot b_c^3}{8 \cdot (1 - v)} \cdot \left[3.2 \cdot \left(\frac{l_c}{b_c}\right) + 0.8\right]$$

$$C_\theta = \rho \cdot c_s \cdot \frac{l_c \cdot b_c^3}{12} \tag{5}$$

$$K = \frac{G \cdot b_c}{2 \cdot (2 - v)} \cdot \left[6.8 \cdot \left(\frac{l_c}{b_c}\right)^{0.65} + 2.4\right]$$

$$C = \rho \cdot c_t \cdot l_c \cdot b_c , \tag{6}$$

where G is soil shear modulus, v is Poisson's ratio, ρ is soil unit density, c_s is soil characteristic wave velocity, for $v < 1/3$ $c_s = c_t \, [0.5(1-2v)(1-v)^{-1}]^{-0.5}$, and for $v > 1/3$ $c_s = 2c_t$, where c_t is soil transversal wave velocity. The reason for using the velocity c_s for rocking motion is explained by Wolf (1994) as follows.

For the rocking motion producing compression and extension, axial waves dominate for small and intermediate values of v, resulting in the use of c_p, which is the longitudinal wave velocity. But c_p tends to infinity for v approaching 0.5. This causes apparently anomalous behaviour. Use of c_p for the higher values of v would overestimate the radiation damping characterized by C_θ. In view of the fact that $c_s = 2c_t$ yields the correct high frequency asymptote of damping for both $v = 1/3$ and $\frac{1}{2}$ and in addition provides a best fit for small frequencies, this value is used throughout the range of nearly incompressible soil (Wolf, 1994).

For use in an explicit algorithm Equations (1), (2), and (3) are reformulated according to Wolf (1994) as:

$$\dot{\theta}_1 = \frac{K_\theta}{3 \cdot C_\theta} \cdot (\theta_o - \theta_1) \tag{7}$$

$$\ddot{\theta}_o = \frac{M - \frac{2}{3} \cdot K_\theta \cdot \theta_o - C_\theta \cdot \dot{\theta}_o - \frac{K_\theta}{3} \cdot \theta_1 + K \cdot e \cdot u_o + C \cdot e \cdot u_o}{I + \Delta M_\theta} \tag{8}$$

$$\ddot{u}_o = \frac{-K \cdot u_o - C_\theta \cdot \dot{u}_o}{m} - e \cdot \ddot{\theta}_o \tag{9}$$

For the parameter θ_1 no prediction or correction are formulated. Starting from the known motion at time $(n-1)\,\Delta t$, that is
θ_{on-1}, $\dot{\theta}_{on-1}$, $\ddot{\theta}_{on-1}$, θ_{1n-1}, $\dot{\theta}_{1n-1}$, u_{on-1}, $\dot{u}_{on-1}$, $\ddot{u}_{on-1}$, the final rotations and displacement and the predicted velocities at time $n\Delta t$ are calculated by the following equations:

$$\theta_{on} = \theta_{on-1} + \Delta t \cdot \dot{\theta}_{on-1} + \frac{\Delta t^2}{2} \cdot \ddot{\theta}_{on-1} \tag{10}$$

$$\theta_{1n} = \theta_{1n-1} + \Delta t \cdot \dot{\theta}_{1n-1} \tag{11}$$

$$u_{on} = u_{on-1} + \Delta t \cdot \dot{u}_{on-1} + \frac{\Delta t^2}{2} \cdot \ddot{u}_{on-1} \tag{12}$$

$$\langle\dot{\theta}\rangle_{on} = \dot{\theta}_{on-1} + \frac{\Delta t}{2} \cdot \ddot{\theta}_{on-1} \tag{13}$$

$$\langle\dot{u}\rangle_{on} = \dot{u}_{on-1} + \frac{\Delta t}{2} \cdot \ddot{u}_{on-1} \tag{14}$$

The symbol $\langle\ \rangle$ denotes a predicted value. Based on these values in place of $\dot{\theta}_{on}$, $\dot{u}_{on}$, the rotational velocity $\dot{\theta}_{1n}$ and the accelerations $\ddot{\theta}_{on}$, $\ddot{u}_{on}$ follow from Equation (6.39), (6.40), and (6.41) formulated at time $n\Delta t$. The two predicted velocities are corrected as

$$\dot{\theta}_{on} = \langle\dot{\theta}\rangle_{on} + \frac{\Delta t}{2} \cdot \ddot{\theta}_{on} \tag{15}$$

$$\dot{u}_{on} = \langle\dot{u}\rangle_{on} + \frac{\Delta t}{2} \cdot \ddot{u}_{on} \tag{16}$$

For stability of the explicit algorithm the time step Δt must be smaller than the (smallest) natural period divided by π that is $2\omega^{-1}$. The rocking natural frequency can be computed according to Wolf (1994):

$$\omega_r = \sqrt{\frac{K_\theta \cdot k_\theta(b_o)}{I + \Delta M_\theta + e^2 \cdot m}} \tag{17}$$

$$k_\theta(b_o) = 1 - \frac{1/3 \cdot b_o^{\,2}}{1 + b_o^{\,2}} \tag{18}$$

$$b_o = \omega_r \cdot \frac{z_o}{c_p} \tag{19}$$

$$z_o = \frac{9 \cdot \pi \cdot r_o}{32} \cdot (1 - v) \cdot \left(\frac{c_s}{c_t}\right)^2 \tag{20}$$

$$r_o = \sqrt[4]{\frac{l_c \cdot b_c^{\,3}}{3 \cdot \pi}} \tag{21}$$

Equation (17) is solved iteratively starting with $\omega_r = 0$. The parameters I, ΔM_θ, K_θ, e, m. l_c, b_c, v, are as in Equations (1) to (3), c_p is the velocity of longitudinal waves through soil beneath wall; c_t is the velocity of transversal waves through soil beneath the wall; $c_s = c_p$ for Poisson's ratio $v < 1/3$ and $c_s = 2\,c_t$ for $v > 1/3$. The circular frequency of horizontal motion is

$$\omega_h = \sqrt{\frac{K}{m}} \tag{22}$$

The fundamental frequency ω of the coupled system can be approximated using the uncoupled natural frequencies according to Wolf (1994)

$$\frac{1}{\omega^2} = \frac{1}{\omega_h^2} + \frac{1}{\omega_r^2} \tag{23}$$

7 Vibration Properties of a Shallow Foundation for Gas Turbine

The work book is referred in Section 7.4.3. A work sheet from the work book is shown in Fig. 7

8 Vibration Properties of a Rubber Bearings Isolated Foundation

The work book is referred in Section 8.2.1.1. A work sheet from the work book is shown in Fig. 8

9 Vibration Properties of a Viscoelastically Damped Foundation

The work book is referred in Section 8.2.2.1. A work sheet from the work book is shown in Fig. 9

10 Vibration Properties of a Passively Isolated Building in Japan – Upper Bound Horizontal Stiffness and Damping Ratio

The work book is referred in Section 8.4.4.1. A work sheet from the work book is shown in Fig. 10

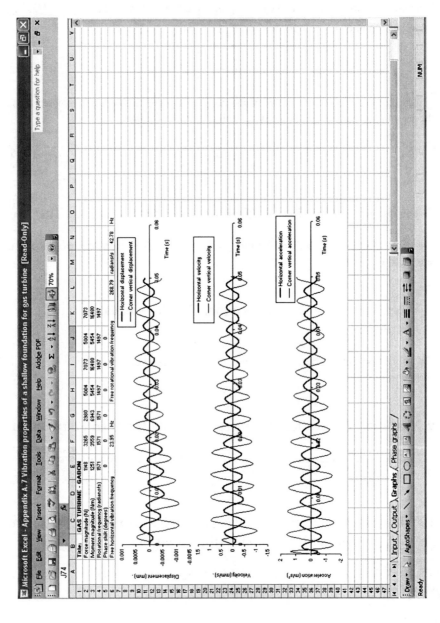

Fig. 7 Work sheet Graph in the work book Appendix 7

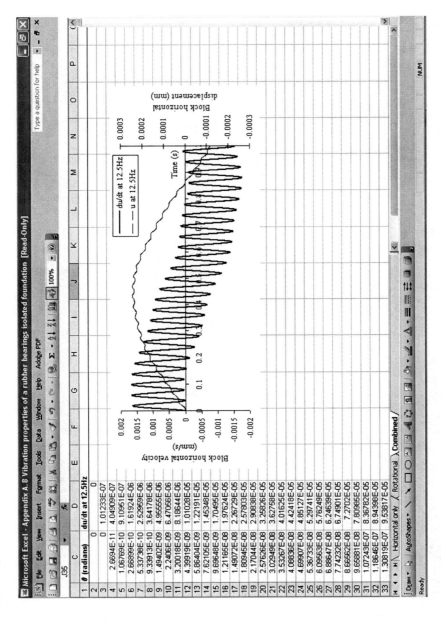

Fig. 8 Work sheet Combined in the work book Appendix 8

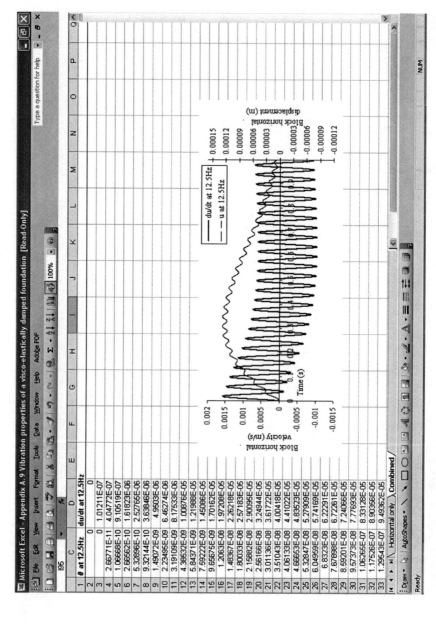

The page shows a Microsoft Excel spreadsheet screenshot (Read Only) titled "Microsoft Excel - Appendix A.9 Vibration properties of a visco-elastically damped foundation [Read Only]" containing a data table and an embedded chart.

Spreadsheet column C header: θ at 12.5Hz
Spreadsheet column D header: du/dt at 12.5Hz

	C (θ at 12.5Hz)	D (du/dt at 12.5Hz)
2	0	0
3	0	1.01211E-07
4		4.04772E-07
5	2.66771E-11	9.10519E-07
6	1.06668E-10	1.61823E-06
7	2.66562E-10	2.52765E-06
8	5.32896E-10	3.63846E-06
9	9.32144E-10	4.9503E-06
10	1.49072E-09	6.46274E-06
11	2.23495E-09	8.17533E-06
12	3.19109E-09	1.00876E-05
13	4.38532E-09	1.21988E-05
14	5.84371E-09	1.45096E-05
15	7.59222E-09	1.70162E-05
16	9.65675E-09	1.97208E-05
17	1.2063E-08	2.26218E-05
18	1.48367E-08	2.57183E-05
19	1.80033E-08	2.90095E-05
20	2.15882E-08	3.24944E-05
21	2.56166E-08	3.61722E-05
22	3.01136E-08	4.00418E-05
23	3.51043E-08	4.41022E-05
24	4.06133E-08	4.83523E-05
25	4.66653E-08	5.27909E-05
26	5.32847E-08	5.74169E-05
27	6.04959E-08	6.22291E-05
28	6.8323E-08	6.72261E-05
29	7.67898E-08	7.24066E-05
30	8.59201E-08	7.77693E-05
31	9.57373E-08	8.33128E-05
32	1.06265E-07	8.90356E-05
33	1.17526E-07	9.49362E-05
	1.29543E-07	

Chart legend:
- du/dt at 12.5Hz
- u at 12.5Hz

Chart left axis: Block horizontal velocity (m/s) — 0.002, 0.0015, 0.001, 0.0005, 0, -0.0005, -0.001, -0.0015
Chart right axis: Block horizontal displacement (m) — 0.00015, 0.00012, 0.00009, 0.00006, 0.00003, 0, -0.00003, -0.00006, -0.00009, -0.00012
Chart x-axis: Time (s)

Fig. 9 Work sheet Combined in the work book Appendix 9

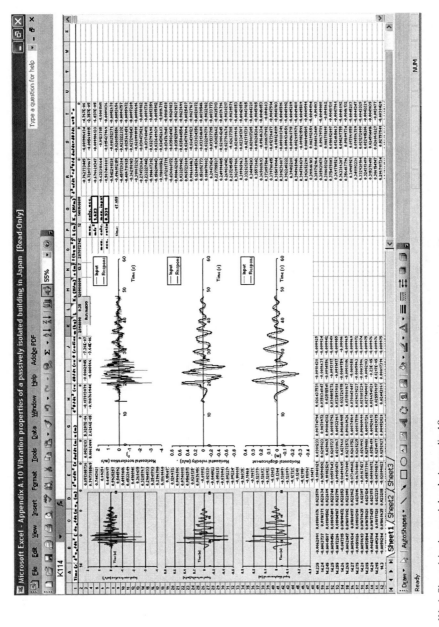

Fig. 10 Work Sheet 1 in the work book Appendix 10

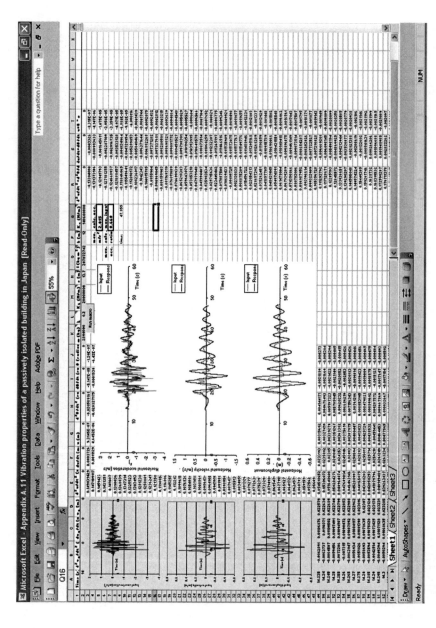

Fig. 11 Work Sheet 1 in the work book Appendix 11

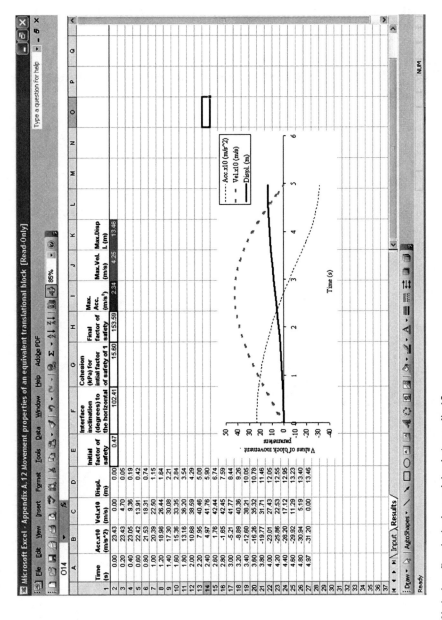

Fig. 12 Work sheet Results in the work book Appendix 12

11 Vibration Properties of a Passively Isolated Building in Japan – Lower Bound Horizontal Stiffness and Damping Ratio

The work book is referred in Section 8.4.4.1. A work sheet from the work book is shown in Fig. 11

12 Fast Movement on Failure of the Asele Road Embankment in Sweden

The work book is referred in Section 9.2.1. A work sheet from the work book is shown in Fig. 12

References

AASHTO (2009) Guide specifications for LRFD Seismic Bridge Design. American Association of State Highway and Transportation Officials, Washington, DC

Ambraseys NN (1988) Engineering seismology. Earthquake Eng Struct Dynam 17:1–105

Ambraseys NN, Hendron AJ (1968) Dynamic behaviour of rock masses. In: Stagg KG, Zienkiewicz OC (eds) Rock mechanics in engineering practice. Wiley, London, pp 203–227

Ambraseys NN, Srbulov M (1995) Earthquake induced displacements of slopes. Soil Dyn Earthquake Eng 14:59–71

Ambraseys NN, Douglas J (2003) Effect of vertical ground motions on horizontal response of structures. Int J Struct Stab Dyn 3:227–266

Ambraseys NN, Douglas J, Sigbjornsson R, Berge-Thierry C, Suhadolc P, Costa G, Smit P (2004) European strong motion database – vol 2. The Engineering and Physical Science Research Council of the United Kingdom GR-52114-01

Amick H (1997) On generic vibration criteria for advanced technology facilities: with a tutorial on vibration data representation. J Inst Environ Sci XL(5):35–44

Arroyo M, Wood DM, Greening PD, Median L, Rio J (2006) Effects of sample size on bender-based axial G0 measurements. Geotechnique 56(1):39–52

ANSI S3.18 (1979) Guide for the evaluation of human exposure to whole-body vibrations. American National Standards Institute, Acoustical Society of America, New York, Secretariat for Committees 51, 52 and 53

ANSI S3.29 (1983) Guide to the evaluation of human exposure to vibration in buildings. American National Standards Institute, Acoustical Society of America, New York, Secretariat for Committees 51, 52 and 53

API RP 2A-WSD (2000) Recommended practice for planning, designing and constructing fixed offshore platforms – working stress design, 21st edn. American Petroleum Institute, Washington, DC

ASTM D1556 – 00 Standard test method for density and unit weight of soil in place by sand-cone method. American Society for Testing and Materials, Annual Book of ASTM Standards 04.08

ASTM D1586 – 08a Standard test method for standard penetration test (SPT) and split-barrel sampling of soils. American Society for Testing and Materials, Annual Book of ASTM Standards 04.08

ASTM D3999 – 91 Standard test methods for the determination of the modulus and damping properties of soils using the cyclic triaxial apparatus. American Society for Testing and Materials, Annual Book of ASTM Standards 04.08

ASTM D4015 – 92 Standard test method for modulus and damping of soils by the resonant-column method. American Society for Testing and Materials, Annual Book of ASTM Standards 04.08

ASTM D4253 Standard test method for maximum index density and unit weight of soils using a vibratory table. American Society for Testing and Materials, Annual Book of ASTM Standards 04.08

ASTM D4254 Standard test method for minimum index density and unit weight of soils and calculation of relative density. American Society for Testing and Materials, Annual Book of ASTM Standards 04.08

ASTM D4428/D 4428 M – 00 Standard test methods for crosshole seismic testing. American Society for Testing and Materials, Annual Book of ASTM Standards 04.08

ASTM D4914 Standard test methods for density of soil and rock in place by the sand replacement method in a trial pit. American Society for Testing and Materials, Annual Book of ASTM Standards 04.08

ASTM D5777 – 00 Standard guide for using the seismic refraction method for subsurface investigation. American Society for Testing and Materials, Annual Book of ASTM Standards 04.08

ASTM D 5778 – 95 (Reapproved 2000) Standard test method for performing electronic friction cone and piezocone penetration testing of soils1. American Society for Testing and Materials, Annual Book of ASTM Standards 04.08

ASTM D6429-99 Standard guide for selecting surface geophysical methods. American Society for Testing and Materials, Annual Book of ASTM Standards 04.08

ASTM D6432-99 Standard guide for using the surface ground penetrating radar method for subsurface investigation. American Society for Testing and Materials, Annual Book of ASTM Standards 04.08

ASTM D7128 – 05 Standard guide for using the seismic-reflection method for shallow subsurface investigation. American Society for Testing and Materials, Annual Book of ASTM Standards

ASTM D7400 – 08 Standard test methods for downhole seismic testing. American Society for Testing and Materials, Annual Book of ASTM Standards

Attewell PB, Farmer IW (1973) Attenuation of ground vibrations from pile driving. Ground Engineering 6(4):26–29

Bahrekazemi M (2004) Train-induced ground vibration and its prediction. PhD thesis, Division of Soil and Rock Mechanics, Department of Civil and Architectural Engineering, Royal Institute of Technology, Stockholm

Baker GL, Gollub JP (1992) Chaotic dynamics, an introduction. Cambridge University Press, Cambridge

Bard PY (1998) Micro tremor measurement: a tool for site effect estimation? Proceedings of the 2nd International symposium on the effects of surface geology on seismic motion, Yokohama, Japan

Barneich JA (1985) Vehicle induced ground motion. In: Gazetas G, Selig ET (eds) Vibration problems in geotechnical engineering, Proceedings of ASCE convention in Detroit, Michigan, pp 187–202

Bell FG, Culshaw MG (1998) Some geohazards caused by soil mineralogy, chemistry and micro fabric: a review. In: Maund JG, Eddleston M (eds) Geohazards in engineering geology. The Geological Society Engineering Geology Special Publication No. 15, London, pp 427–442

Benioff H (1934) The physical evaluation of seismic destructiveness. Bull Seismol Soc Am 24:398–403

Biot MA (1941) A mechanical analyzer for the prediction of earthquake stress. Bull Seismol Soc Am 31:151–171

Bishop AW (1955) The use of slip circle for stability analysis. Geotechnique 5(1):7–17

Bommer J (1992) The recording, interpretation and use of earthquake strong-motion. Internal report. Imperial College of Science, Technology and Medicine, London

Bormann P (ed) (2002) New manual of seismological observatory practice. GeoForschungsZentrum, Potsdam

Boulanger RW, Idriss IM (2007) Evaluation of cyclic softening in silt and clays. J Geotech Geoenviron Eng, ASCE 133:641–652

Boussinesq J (1885) Application des potentials a l'etude de l'equilibre et du movement des solides elastique. Gauthier-Villard, Paris

BS 1377-2 (1990) Soils for civil engineering purposes. Part 2: Classification tests. British Standard Institution, London

BS 1377-4 (1990) Soils for civil engineering purposes. Part 4: Compaction related tests. British Standard Institution, London

BS 5228-2 (2009) Code of practice for noise and vibration control on construction and open sites – Part 2: Vibration. British Standard Institution, London

BS 6472 (1992) Guide to evaluation of human exposure to vibration in buildings (1 Hz to 80 Hz). British Standard Institution, London

BS 6955-3 (1994) Calibration of vibration and shock pick-ups. Part 3. Method for secondary vibration calibration. British Standard Institution, London

BS 7385-2 (1993) Evaluation and measurement for vibration in buildings, Part 2: Guide to damage levels from ground borne vibration. British Standard Institution, London

Burland JB (1989) Small is beautiful – the stiffness of soils at small strains. Can Geotech J 26:499–516

Caltrans (2001) Ground vibration monitoring for construction blasting in urban areas, Report F-00-OR-10. State of California Department of Transportation, Sacramento, CA

Cetin KO, Seed RB, Kiureghian AD, Tokimatsu K, Harder LH Jr, Kayen RE, Moss RES (2004) Standard penetration test-based probabilistic and deterministic assessment of seismic soil liquefaction potential. J Geotech Geoenviron Eng, ASCE 130:1314–1340

Charlie WA, Jacobs A, Doehring DO (1992) Blast induced liquefaction of an alluvial sand deposit. Geotech Test J 15(1), 14–23

Chatfield C (1992) The analysis of time series – an introduction, 4th edn. Chapman & Hall, London

Clayton CRI, Priest JA, Bui M, Zervos A, Kim SG (2009) The Stokoe resonant column apparatus: effects of stiffness, mass and specimen fixity. Geotechnique 59(5):429–437

Clough GW, Chameau JL (1980) Measured effects of vibratory sheep pile driving. J Geotech Eng Div, ASCE 104(GT10):1081–1099

Clough RW, Penzien J (1993) Dynamics of structures, 2nd edn. McGraw-Hill, New York, NY

Cooley JW, Tukey JW (1965) An algorithm for machine calculation of complex Fourier series. Mathematics of computing, reprinted 1972. In: Rabiner LR, Rader CM (eds) Digital signal processing. IEEE Press, New York, NY, pp 223–227

Cornforth DH (1964) Some experiments on the influence of strain conditions on the strength of sand. Geotechnique 16(2):143–167

CP 2012-1 (1974) Code of practice for foundations for machinery – part 1: foundations for reciprocating machines. British Standards Institution, London

Das BM (1985) Advanced soil mechanics. Hemisphere Publishing Corporation a McGraw-Hill, New York, NY

Dawn TM, Stanworth CG (1979) Ground vibration from passing trains. J Sound Vib 66(3):355–362

Dean ETR (2009) Offshore geotechnical engineering – principles and practice. Thomas Telford, London

DIN 4024-1 (1988) Maschinenfundamente; Elastische Stützkonstruktionen für Maschinen mit rotierenden Massen

DIN 4024-2 (1991) Maschinenfundamente; Steife (starre) Stützkonstruktionen für Maschinen mit periodischer Erregung

DIN 4150-3 (1999) Erschütterungen im Bauwesen – Teil 3: Einwirkungen auf bauliche Anlage Norm Ausgabe, Deutsch, Bestellen beim Beuth Verlag

Douglas J (2003) What is a poor quality strong-motion record? Bull Earthquake Eng 1:141–156

Dowding CH (2000) Construction vibration. Reprinted 1996 version. Prentice Hall, Englewood Cliffs, NJ

Dyvik R, Madshus C (1985) Lab measurements of G_{max} using bender elements. In: Khosla V (ed) Advances in the art of testing soils under cyclic conditions. Proceedings of Geotechnical Engineering Division of ASCE Convention in Detroit, Michigan, pp 186–196

Eurocode 7: Geotechnical design. Part 2: Ground investigation and testing 2007, EN 1997-2:2007. European Committee for Standardization (CEN)

Eurocode 8: Design of structures for earthquake resistance – Part 1: General rules, seismic actions and rules for buildings, EN1998-1:2004. European Committee for Standardization (CEN)

Eurocode 8: Design of structures for earthquake resistance – Part 5: Foundations, retaining structures and geotechnical aspects, EN1998-1:2004. European Committee for Standardization (CEN)

Ekstrom A, Olofsson T (1985) Water and frost – stability risks for embankments of fine grained soils. In: Symposium on failures in earthworks. Institution of Civil Engineers, London, pp 155–166

Eldred PJL, Skipp BP (1998) Vibration on impact. In: Skipp BO (ed) Ground dynamics and man-made processes. Institution of Civil Engineers, London

Finn WDL (1985) Aspects of constant volume cyclic simple shear. In: Khosla V (ed) Advances in the art of testing soils under cyclic conditions. Proceedings of Geotechnical Engineering Division of ASCE Convention in Detroit, Michigan, pp 74–98

Fornaro M, Patrucco M, Sambuelli L (1994) Vibrations from explosives, high energy hydraulic hammers and TBMs; experience from Italian tunnels. Gallerie e Grandi Opere Sotterranee, December 30–37

Forsblad L (1974) Markskakningar och deras skadeverkan, inverkan av vibrerande jordpack-ningamaskiner. Skadlig Inverkan av Vibrationer. Sympsium Anordnat av Svenska Geotekniska Foreningen rapporter 56

Gandhi SR, Dey AK, Selvam S (1999) Densification of pond ash by blasting. J Geotech Geoenviron Eng 125(10):889–899

Gazetas G (1991) Foundation vibrations. In: Fang H-Y (ed) Foundation engineering handbook (2nd edn), Chapman & Hall, London, pp 553–593

Gerrard CM (1977) Background to mathematical modelling in geomechanics: the roles of fabric and stress history. In: Gudehus G (ed) Finite elements in geomechanics. Wiley, New York, NY, pp 33–120

Gieck K, Gieck R (1997) Engineering formulas, 7th edn. McGraw-Hill, New York, NY

Gohl WB, Jefferies MG, Howie JA (2000) Explosive compaction: design, implementation and effectiveness. Geotechnique 50:6576–665

Gregory CE (1984) Explosives for North American engineers, 3rd ed. Trans Tech Publications, Uetikon-Zuerich

Hall JR (ed) (1987) Use of vibration measurement in structural evaluation. Proceedings of a session sponsored by the Structural Division of the American Society of Civil Engineers in conjunction with the ASCE convention in Atlantic City, New Jersey

Hamaguchi H, Higashino M, Iiba M, Kani N, Kasai K, Midorikawa M (2006) Overview of response controlled buildings. In: Higashino M, Okamoto S (eds) Response control and seismic isolation of buildings. Taylor & Francis, London and New York, NY, pp 207–217

Hanna TH (1985) Field instrumentations in geotechnical engineering. Trans Tech Publications, Uetikon-Zuerich

Hardin BO (1978) The nature of stress-strain behavior of soil. In: Earthquake engineering and soil dynamics. ASCE, Pasadena, CA, 1:3–89

Hardin BO, Drnevich VP (1972) Shear modulus and damping in soil: design equations and curves. J Soil Mech Found Div, ASCE 98:667–692

Hashiguchi K (2001) Description of inherent/induced anisotropy of soils: rotational hardening rule with objectivity. Soils Found 41:139–146

Head JM, Jardine FM (1992) Ground-borne vibrations arising from piling. Construction Industry Research and Information Association, London, Technical Note 142

Heckman WS, Hagerty DJ (1978) Vibrations associated with pile driving. J Constr Div, ASCE 104(CO4):14205

Hiller DM, Crabb GI (2000) Groundborne vibration caused by mechanised construction works. Transport Research Laboratory Report 429, United Kingdom

Hoek E, Brown ET (1980) Underground excavations in rock. The Institution of Mining and Metallurgy, London

Hoek E, Bray J (1981) Rock slope engineering, revised 3rd ed. The Institution of Mining and Metallurgy, London

Holm G, Andreasson B, Bengtsson PE, Bodare A, Eriklsson H (2002) Mitigation of track and ground vibrations by high speed trains at Ledsgard, Sweden. Swedish Deep Stabilization Research Centre Report 10

HSE (2005) Whole-body vibration – the control of vibration at work regulations. Health and Safety Executive, Suffolk

Hryciw RD (1986) A study of the physical and chemical aspects of blast densification of sand. PhD thesis, Northwestern University, Evanston, IL

Hryciw RD, Vitton S, Thomann TG (1990) Liquefaction and flow failure during seismic exploration. J Geotech Eng Am Soc Civil Eng 116:1881–1899

Huang CJ, Yin HY, Chen CY, Yeh CH, Wang CL (2007) Ground vibrations produced by rock motions and debris flows. J Geophys Res 112

Hungr O, Morgenstern NR (1984) High velocity ring shear test on sand. Geotechnique 34:415–421

Ishibashi I (1992) Discussion to Effect of soil plasticity on cyclic response by M.Vucetic and R.Dobry. J Geotech Eng, ASCE 118:830–832

Ishihara K, Nagase H (1985) Multi-directional irregular loading tests on sand. In: Khosla V (ed) Advances in the art of testing soils under cyclic conditions. Proceedings of Geotechnical Engineering Division of ASCE convention in Detroit, Michigan, pp 99–119

ISO 4866 (1990) Evaluation and measurement for vibration in buildings. Part 1: Guide for measurement of vibrations and evaluation of their effects on buildings. International Organization for Standardization, Geneva

ISO 16063-16 (2003) Methods for the calibration of vibration and shock transducers – part 11: Primary vibration calibration by laser interferometry. International Organization for Standardization, Geneva

ISO 16063-21 (2003) Methods for the calibration of vibration and shock transducers – part 21: Vibration calibration by comparison to a reference transducer. International Organization for Standardization, Geneva

ISO 16063-22 (2003) Methods for the calibration of vibration and shock transducers – part 22: Shock calibration by comparison to a reference transducer. International Organization for Standardization, Geneva

Itoh K, Zeng X, Koda M, Murata O, Kusakabe O (2005) Centrifuge simulation of wave propagation due to vertical vibration on shallow foundations and vibration attenuation countermeasures. J Vib Control 11(6):781–800

Ivanovic SS, Trifunac MD, Novikova EI, Gladkov AA, Todorovska MI (2000) Ambient vibration tests of a seven-story reinforced concrete building in Van Nuys, California, damaged by the 1994 Northridge earthquake. Soil Dyn Earthquake Eng 19:391–411

Jaksa MB, Griffith MC, Grounds RW (2002) Ground vibrations associated with installing enlarged-base driven cast-in-situ piles. Aust Geomech 37:67–73

Jaky J (1944) The coefficient of earth pressure at rest. Magyar Mernok es Epitez Egylet Kozlonye

Japanese Ministry of Transport (1999) Handbook on liquefaction remediation of reclaimed land. AA. Balkema, Rotterdam

Japan Road Association (2003) Specification for highway bridges, part V – seismic design. English version translated by PWRI, Japan. Japan Road Association, Tokyo

Jardine RJ, Fourier AB, Maswoswse J, Burland JB (1985) Field and laboratory measurements of soil stiffness. In: Proceedings of the 11th international conference on soil mechanics and foundation engineering, San Francisco, CA, vol 2, pp 511–514

Jefferies MG, Been K (2006) Soil liquefaction. Taylor & Francis, London

Jenkins WM (1989) Theory of structures. Chapter 3. In: Blake LS (ed.) Civil Engineer's Reference Book, 4th ed., Butterworths, pp 3–16

Kahriman A (2004) Analysis of parameters of ground vibration produced from bench blasting at a limestone quarry. Soil Dyn Earthquake Eng 24:887–892

Kaino T, Kikuchi T (1988) Earthquake response of pier with caisson type wall foundation and its analysis. Proceedings of the 9th world conference on earthquake engineering, Tokyo, vol 3, 5–4–16

Kramer SL (1996) Geotechnical earthquake engineering. Prentice Hall, Englewood Cliffs, NJ

Kogut J, Degrande G, Lombaert G, Pyl L (2004) Measurements and numerical prediction of high speed train vibrations. Proceedings of the 4th International Conference on Case Histories in Geotechnical Engineering, New York, Paper No. 4.02

Konrad JM, Watts BD (1995) Undrained shear strength for liquefaction flow failure analysis. Can Geotech J 32:783–794

Lacy HS, Gould JP (1985) Settlement from pile driving in sands. In: Gazetas G, Selig ET (eds) Vibration problems in geotechnical engineering, Proceedings of ASCE convention in Detroit, Michigan, pp 152–173

Ladd CC, Foot R (1974) New design procedures for stability of soft clays. J Geotech Eng Div, ASCE 100:763–786

Lambe TW, Whitman RV (1979) Soil mechanics, SI version. Wiley, New York, NY

Langhaar HL (1951) Dimensionless analysis and theory of models. Wiley, New York, NY

Lee KL, Focht JA (1976) Strength of clay subjected to cyclic loading. Marine Geotechnol 1:305–325

Lee J-S, Sanatamarina JC (2005) Bender elements: performance and signal interpretation. J Geotech Geoenviron Eng, ASCE 131(9):1063–1070

LESSLOSS (2007) Innovative anti-seismic systems user manual. In: Forni M (ed) LESSLOSS – Risk mitigation for earthquakes and landslides, Report No. 2007/03 http://www.lessloss.org/main/index.php

Liao SSC, Whitman RV (1986) Overburden correction factors for SPT in sand. J Geotech Eng, ASCE 112:373–377

Linehan PW, Longinow A, Dowding CH (1992) Pipeline response to pile driving and adjacent excavation. J Geotech Div, ASCE 118(2):300–316

List BR, Lord ERF, Fair AE (1985) Investigation of potential detrimental vibrational effects resulting from blasting in oilsands. In: Gazetas G, Selig ET (eds) Vibration problems in geotechnical engineering. Proceedings of a symposium of Geotechnical Engineering Division of ASCE, Detroit, Michigan, pp 266–285

Lucca FJ (2003) Tight construction blasting: ground vibration basics, monitoring, and prediction. Terra Dinamica L.L.C., Granby, CT

Lunne T, Robertson PK, Powell JJM (2001) Cone penetration testing in geotechnical practice. Spon Press, London

Maksimovic M (1988) General slope stability software package for micro computers. Proceedings of the 6th international conference on numerical methods in geomechanics, Innsbruck 3:2145–2150

Martin DJ (1980) Ground vibrations from impact pile driving during road construction. U.K. Transport and Road Research Laboratory Report 544

Mayne PW (1985) Ground vibrations during dynamic compaction. In: Gazetas G, Selig ET (eds) Vibrations problems in geotechnical engineering. Proceedings of ASCE convention in Detroit, Michigan, pp 247–265

McDowell PW, Barker RD, Butcher AP, Culshaw MG, Jackson PD, McCann DM, Skip BO, Matthews SL, Arthur JCR (2002) Geophysics in engineering investigations. Report C562 of Construction Industry Research and Information Association, London

McLelland Engineers (1977) extract from technical report. In: Poulos HG (1988) Marine Geotechnics. Unwin Hyman, Boston, MA

Miller Engineers Inc (1992) Vibration monitoring at Manchester airport, Manchester, NH

Mitchell JK (1993) Fundamentals of soil behaviour. Wiley, New York, NY

Mitchell JK, Houston WN (1969) Causes of clay sensitivity. J Soil Mech Found Div, ASCE 95(3):845–869

Mooney MA, Rinehart RV (2007) Field monitoring of roller vibration during compaction of subgrade soil. J Geotech Geoenviron Eng, ASCE 133(3):257–265

Nabeshima Y, Hayakawa K, Kani Y (2004) Experimental and numerical studies on ground vibration isolation by PC wall-pile barrier. Proceedings of the 5th International Conference on Case Histories in Geotechnical Engineering, New York, Paper No. 4.06

Nakamura Y (1989) A method for dynamic characteristics estimation of subsurface using micro tremor on the ground surface. Q Rep Railway Tech Res Inst 30(1)

Neilson PP, O'Rourke TD, Flanagan RF, Kulhaway FH, Ingraffea AR (1984) Tunnel boring machine performance study. U.S. Department of Transportation Report UMTA-MA-06-0100-84-1

New BM (1986) Ground vibration caused by civil engineering works. U.K. Transport and Road Research Laboratory Research Report 53

Newmark NM, Hall WJ, Morgan JR (1977) Comparison of building response and free field motion in earthquakes. The 6th world conference on earthquake engineering, New Delhi, vol 2, pp 972–977

Nogoshi M, Igarashi T (1971) On the amplitude characteristics of microtremor – part 2. J Seismol Soc Japan, 24:26–40 (in Japanese)

Novak M, Grigg RF (1976) Dynamic experiments with small pile foundation. Can Geotech J 13:372–395

Olson SM (2001) Liquefaction analysis of level and sloping ground using field case histories and penetration resistance. PhD thesis, University of Illinois at Urbana-Champaign, IL

Olson SM, Stark TD (2002) Liquefied strength ratio from liquefaction flow failure case histories. Can Geotech J 39:629–647

Oriard LL, Richardson TL, Akins KP (1985) Observed high rise building response to construction blast vibration. In: Gazetas G, Selig ET (eds) Vibration problems in geotechnical engineering. Proceedings of a symposium of Geotechnical Engineering Division of ASCE, Detroit, Michigan, pp 203–228

Paine JG (2003) Assessing vibration susceptibility over shallow and deep bedrock using accelerometers and walkway surveys. Proceedings of symposium on the application of geophysics to engineering and environmental problems, Environmental and Engineering Geophysical Society, pp 1263–1275 (CD-ROM)

Parathiras A (1995) Rate of displacement effects on fast residual strength. In: Ishihara K (ed) The 1st international conference on earthquake geotechnical engineering, Tokyo, vol 1, pp 233–237

Peacock WH, Seed HB (1968) Sand liquefaction under cyclic loading simple shear conditions. J Soil Mech Found Div, ASCE 94(SM3):689–708

Peck RB, Hanson WE, Thornburn TH (1974) Foundation engineering, 2nd ed. Wiley, New York, NY

Potts DM, Zdravkovic L (1999) Finite element analysis in geotechnical engineering – theory. Thomas Telford, London

Poulos HG (1979) Group factors for pile deflection estimation. J Geotech Eng Div, ASCE 105:1489–1509

Rankka K, Andersson-Skold I, Hulten K, Larson R, Leroux V, Dahlin T (2004) Quick clay in Sweden. Swedish Geotechnical Institute Report 65

Richart FE, Hall JR, Woods RD (1970) Vibrations of soils and foundations. Prentice-Hall, Englewoods Cliffs, NJ

Rivin EI (2003) Passive vibration isolation. The American Society of Mechanical Engineers and Professional Engineering Publishing Limited, London

Rix GJ, Stokoe KH (1992) Correlation of initial tangent modulus and cone resistance. Proceedings of the international symposium on calibration chamber testing, Potsdam, New York, 1991. Elsevier, Amsterdam, pp 351–362

Robertson PK, Campanella RG, Gillespie D, Greig J (1986) Use of piezometer cone data. Proceedings of the ASCE specialty conference in situ '86: Use of in situ tests in geotechnical engineering, American Society of Civil Engineers, Blacksburg, pp 1263–1280

Roscoe KH (1953) An apparatus for the application of simple shear to soil samples. Proceedings of the 3rd international conference on soil mechanics, Zurich, vol 1, pp 186–191

Sarma SK, Srbulov M (1996) A simplified method for prediction of kinematic soil-foundation interaction effects on peak horizontal acceleration of a rigid foundation. Earthquake Eng Struct Dyn 25:815–836

Sarma SK, Srbulov M (1998) A uniform estimation of some basic ground motion parameters. J Earthquake Eng 2(2):267–287

Schaefer VR (ed), Ambramson LW, Drumheller JC, Hussin JD, Sharp KD (1997) Ground improvement, ground reinforcement, ground treatment developments 1987–1997. ASCE Geotechnical Special Publication 69

Seed HB (1979) Soil liquefaction and cyclic mobility evaluation for level ground during earthquakes. J Geotech Eng Div, ASCE 105(GT2):201–255

Seed HB, Chan CK (1959) Thixotropic characteristics of compacted clay. Trans Am Soc Civil Eng 124:894–916

Seed HB, Idriss IM (1967) Analysis of soil liquefaction: Niigata earthquake. J Soil Mech Found Div, ASCE 93(SM3):83–108

Seed HB, Idriss IM (1970) Soil modules and damping factors for dynamic response analyses. Report EERC 70-10, Earthquake Engineering Research Center, University of California, Berkeley, CA

Seed HB, Idriss IM, Makdisi F, Banerjee N (1975) Representation of irregular stress time histories by equivalent uniform stress series in liquefaction analyses. Report EERC 75-29, Earthquake Engineering Research Center, University of California, Berkeley, CA

Seed HB, Tokimatsu K, Harder LF, Chung RM (1985) Influence of SPT procedures in soil liquefaction resistance evaluations. J Geotech Eng, ASCE 111:1425–144

Shakal AF, Huang MJ, Darragh RB (1996) Interpretation of significant ground response and structure strong motions recorded during the 1994 Northridge earthquake. Bull Seismol Soc Am 86:S231–S246

Sheng X, Jones CJC , Thompson DJ (2006) Prediction of ground vibration from trains using the wave number finite and boundary element methods. J Sound Vib 293(3–5):575–586

Shirley DJ (1978) An improved shear wave transducer. J Acoust Soc Am 63(5):1643–1645

Simsek O, Dalgic S (1997) Consolidation properties of the clays at Duzce plain and their relationship with geological evolution. Geol Bull Turkey 40(2):29-38

Skempton AW (1957) Discussion: The planning and design of new Hong Kong airport. Proc Inst Civil Eng 7:305–307

Skempton AW (1986) Standard penetration test procedures and the effects in sands of overburden pressure, relative density, particle size, aging and over consolidation. Geotechnique 36:425–447

Skipp BO (1998) Ground vibration – codes and standards. In: BO Skipp (ed) Ground dynamics and man-made processes. The Institution of Civil Engineers, Birmingham

Skoglund GR, Marcuson WF3rd , Cunny RW (1976) Evaluation of resonant column test devices. J Geotech Eng Div, ASCE 11:1147–1158

SNAME (1997) Site specific assessment of mobile jack-up units, 1st edn with revision 1. The Society of Naval Architects and Marine Engineers, Jersey City, NJ, pp 61–69

Soong TT, Dargush GF (1997) Passive energy dissipation systems in structural engineering. Wiley, Chichester

Srbulov M (2008) Geotechnical earthquake engineering – simplified analyses with case studies and examples. Springer, Berlin

Stroud MA (1988) The standard penetration test – its application and prediction. The Proceedings of the geotechnology conference 'penetration testing in the UK' organized by the Institution of Civil Engineers in Birmingham

Svinkin MR (2002) Predicting soil and structure vibrations from impact machines. J Geotech Geoenviron Eng, ASCE 128(7):602–612

Tajimi H (1984) Predicted and measured vibrational characteristics of a large-scale shaking table foundation. Proceedings of the 8th world conference on earthquake engineering, San Francisco, CA, 3:873–880

Takemiya H (2004) Field vibration mitigation by honeycomb WIB for pile foundations of a high-speed train viaduct. Soil Dyn Earthquake Eng 24:69–87

Tallavo F, Cascante G, Pandey M (2009) Experimental and numerical analyses of MASW tests for detection of buried timber trestles. Soil Dyn Earthquake Eng 29:91–102

Tatsuoka F, Jardine RJ, Presti DLo, Benedetto HDi, Kodaka T (1997) Theme lecture: characterizing the pre-failure deformation properties of geomaterials. Proceedings of the 14th international conference on soil mechanics and foundation engineering, vol 4, pp 2129–2163

Tchepak S (1986) Design and construction aspects of enlarged base Franki piles. Specialty geomechanics symposium, Adelaide, SA, pp 160–165

Ter-Stepanian G (2000) Quick clay landslides: their enigmatic features and mechanism. Bull Eng Geol Environ 59:47–57

Terzaghi K, Peck RB (1948) Soil mechanics in engineering practice. Wiley, New York, NY

Thompson WT (1965) Vibration theory and application. Prentice Hall, Englewood Cliffs, NJ, pp 43–44

Timoshenko SP, Goodier JN (1970) Theory of elasticity, 3rd edn. McGraw-Hill, New York, NY

Tomlinson MJ (2001) Foundation design and construction – 6th edn. Chapman & Hall, London

Towhata I (2007) Developments of soil improvement technologies for mitigation of liquefaction risk. In: Pitilakis KD (ed) Earthquake geotechnical engineering. Springer, Berlin

Tromans I (2004) Behaviour of buried water supply pipelines in earthquake zones. PhD thesis, Imperial College, University of London, London. http://www3.imperial.ac.uk/pls/portallive/docs/1/965899.PDF

Townsend FC (1978) A review of factors affecting cyclic triaxial tests. Special Technical Publication 654, ASTM 356–358

U.S. Bureau of Mines (1971) Blasting vibrations and their effects on structures. Bulletin 656, by Nicholiss HR, Johnson CF, Duvall WI

U.S.-D.O.T. (1995) Transit noise and vibration impact assessment. United States Department of Transport Report DOT-T-95-16, Washington, DC

U.S.-D.O.T. (1998) High-speed ground transportation noise and vibration impact assessment. Office of Railroad Development. United States Department of Transport Report 293630-1, Washington, DC

USBM RI 8507 (1980) by Siskind DE, Stagg MS, Kopp JW, Dowding CH, Structure response and damage produced by ground vibrations from surface blasting. U.S. Bureau of Mines, Washington, DC

Vanmarke EH (1976) Structural response to earthquakes. In: Lomnitz C, Rosenblueth E (eds) Seismic risk and engineering decisions. Elsevier, Amsterdam pp 287–338

Van Impe WF (1989) Soil improvement techniques and their evolution. Balkema, Rotterdam

Verruijt A (1994) Soil dynamics. Delft University of Technology, Delft

Vucetic M, Dobry R (1991) Effect of soil plasticity on cyclic response. J Geotech Eng, ASCE 117:89–107

Watts GR (1987) Traffic-induced ground-borne vibrations in dwellings. UK Transport and Road Research Laboratory Report 102

White D, Finlay T, Bolton MD, Bearss G (2002) Press-in piling: Ground vibration and noise during pile installation. Proceedings of the international deep foundations congress, Orlando, USA. ASCE Special Publication 116:363–371

Wiss JF (1981) Construction vibrations: state of the art. J Geotech Div ASCE 94 (9):167–181

Wolf JP (1994) Foundation vibration analysis using simple physical models. PTR Prentice Hall, Englewood Cliffs, NJ

Wolf JP, Deeks AJ (2004) Foundation vibration analysis: a strength-of-materials approach. Elsevier, Amsterdam

Wood WL (1990) Practical time stepping schemes. Clarendon Press, Oxford.

Woods RD, Jedele LP (1985) Energy – attenuation relationships from construction vibrations. In: Gazetas G, Selig ET (eds) Vibration problems in geotechnical engineering, Proceedings of ASCE convention in Detroit, Michigan, pp 229–246

Yashima A, Oka F, Konrad JM, Uzuoka R, Taguchi Y (1997) Analysis of a progressive flow failure in an embankment of compacted till. Proceedings on Deformation and Progressive Failure in Geomechanics, Japan, Nagoya, vol 1, pp 599–604

Youd TL, Idriss IM (2001) Liquefaction resistance of soil. Summary report from the 1996 NCEER and 1998 NCEER/NSF workshops on evaluation of liquefaction resistance of soils. J Geotech Geoenviron Eng, ASCE 127:297–3137

Zhang J, Andrus RD, Juang CH (2005) Normalized shear modulus and material damping ratio relationships. J Geotech Geoenviron Eng, ASCE 131:453–464

Zienkiewich OC, Taylor RL (1991) The finite element method, 4th edn, vol 2. McGraw Hill, London

Index

Lightning Source UK Ltd.
Milton Keynes UK
29 July 2010

157465UK00005B/77/P

9 789048 190812